▶ Preparing for Life in Humanity 2.0

DOI: 10.1057/9781137277077

Also by Steve Fuller

SOCIAL EPISTEMOLOGY

PHILOSOPHY OF SCIENCE AND ITS DISCONTENTS

PHILOSOPHY, RHETORIC AND THE END OF KNOWLEDGE

SCIENCE (CONCEPTS IN THE SOCIAL SCIENCES)

THE GOVERNANCE OF SCIENCE

THOMAS KUHN: A Philosophical History for Our Times

KNOWLEDGE MANAGEMENT FOUNDATIONS

KUHN VS. POPPER: The Struggle for the Soul of Science

THE INTELLECTUAL

THE NEW SOCIOLOGICAL IMAGINATION

THE PHILOSOPHY OF SCIENCE AND TECHNOLOGY STUDIES

THE KNOWLEDGE BOOK: Key Concepts in Philosophy,
Science and Culture

NEW FRONTIERS IN SCIENCE AND TECHNOLOGY STUDIES

SCIENCE VS. RELIGION?: Intelligent Design and the Problem of
Evolution

DISSENT OVER DESCENT: Intelligent Design's Challenge to
Darwinism

THE SOCIOLOGY OF INTELLECTUAL LIFE

SCIENCE (THE ART OF LIVING)

HUMANITY 2.0

DOI: 10.1057/9781137277077

palgrave▸pivot

Preparing for Life in Humanity 2.0

Steve Fuller

University of Warwick, UK

palgrave
macmillan

DOI: 10.1057/9781137277077

First published 2013 by
PALGRAVE MACMILLAN

Palgrave Macmillan in the UK is an imprint of Macmillan Publishers Limited,
registered in England, company number 785998, of Houndmills, Basingstoke,
Hampshire RG21 6XS.

Palgrave Macmillan in the US is a division of St Martin's Press LLC,
175 Fifth Avenue, New York, NY 10010.

Palgrave Macmillan is the global academic imprint of the above companies
and has companies and representatives throughout the world.

Palgrave® and Macmillan® are registered trademarks in the United States,
the United Kingdom, Europe and other countries.

ISBN: 978-1-137-27708-4 EPUB
ISBN: 978-1-137-27707-7 PDF
ISBN: 978-1-137-27706-0 Hardback

A catalogue record for this book is available from the British Library.

A catalog record for this book is available from the Library of Congress.

www.palgrave.com/pivot

DOI: 10.1057/9781137277077

Contents

DOI: 10.1057/9781137277077

palgrave▸**pivot**

www.palgrave.com/pivot

DOI: 10.1057/9781137277077

Introduction

How do you prepare for life in a state-of-being that might be reasonably called 'Humanity 2.0'? The question was bound to be asked, given the aspirations of 'Humanity 1.0', my name for the sort of being that laws in contemporary democratic societies are designed to empower and protect – namely, someone licensed to explore his or her capacities but without preventing others from doing likewise. It is common to speak of such societies as *liberal* but I believe that *republican* is more appropriate because republicanism implies a sense of collective identity, the definition, protection and possible extension of which provides the common material and ideational basis for the relevant individuals – typically called 'citizens' – to pursue their individually diverse lives (cf. Fuller 2000a: chap. 1). Whereas the protection of liberal freedom tends to target the sort of mutual interference normally associated with crime and discrimination, the protection of republican freedom is more focused on threats to the security of the citizenry as a whole, such as natural disasters, resource scarcity, foreign aggression and epidemics. But why fuss over this nuance in political theory?

Republicanism's relevance to Humanity 2.0 lies in its overriding concern with the construction and maintenance of boundaries – where to draw the line between 'us' and 'them'. When the Weimar-turned-Nazi jurist Carl Schmitt (1996) famously argued that the fundamental problem of politics is determining friend from foe, he was speaking

Fuller, Steve. *Preparing for Life in Humanity 2.0.* Basingstoke: Palgrave Macmillan, 2013.
DOI: 10.1057/9781137277077.

as a republican in this deep sense. Modern democratic societies have found it relatively easy to address the republican question by presuming that all potential citizens are born *Homo sapiens*, even if not native to a given republic's jurisdiction. In that case, non-natives may become 'naturalised' citizens by demonstrating the competence needed to thrive and contribute to the life of the republic. But in Humanity 2.0, that first step – the necessity (and maybe sufficiency) of being born *Homo sapiens* – is removed. Some animals and androids may come to be eligible for citizenship, while some humans may lose their citizenship, perhaps even in the course of their own lifetime. The point of this book is to prepare you intellectually for dealing with this situation.

In the book entitled *Humanity 2.0* (Fuller 2011), I argued that for the past half-century, humanity's self-understanding has been pulled in two opposing directions: the first, promoted by both ecology and evolutionary theory, is towards our greater reembedding in the natural environment; the second, which ultimately aspires to a digital incarnation of humanity, aims for the enhancement, if not outright replacement, of the bodies of our birth. In the former case, we are animals with only a temporary edge over the rest of nature, in the latter we are beings with the capacity to surpass all other animals indefinitely, in which case 'nature' simply refers to the set of obstacles in the way of realising that mission. But in whichever direction Humanity 2.0 turns, the biological species *Homo sapiens* is losing its salience as the default setting for organising the human condition. In the future, it may be seen as a rough draft for some other form of being that we care to dignify as 'human'.

If this conclusion seems paradoxical, consider the meaning of 'genetics' today. The word was coined a century ago by William Bateson to capture the 'science of heredity', but that was before we knew how to manipulate genes for purposes of altering, let alone constructing, individual life-forms. As we have increasingly mastered these skills (with the real scientific-step change occurring with the discovery of the double helix structure of DNA), the sense of 'genetic' as imposing an historic 'load' or 'burden' on future life-forms has receded from view, leaving a more basic sense of 'genetic' as 'building block' or 'raw material', which makes the proposed neologism 'biocapital' for what used to be called 'gene pool' an apposite one (Rose 2006). This point would be more widely appreciated were it not for popular Darwinian accounts of evolution that continue to speak of, say, our 'reptilian brains' as somehow holding our minds in primitive gridlock. Indeed, on the contrary, we are entering a time

DOI: 10.1057/9781137277077

when we have never been 'freer' in the sense that is crucial for moral autonomy, namely, being knowingly responsible for our decisions. Of course, not everything will go to plan as beings are built and rebuilt to fit the world in which we would like them and us to live. However, any future disasters related to these world-building schemes are unlikely to be attributed to the distant past somehow managing to act upon us, what systems theorists call 'hysteresis' (Elster 1976). If anything, more use will be made of the old Jesuitical distinction between *intended* and *anticipated* consequences, which over the past 500 years has been typecast as 'the end justifies the means'. The Jesuits originally argued that the divine signature of nature's overall intelligent design includes locally bad outcomes that the deity has anticipated but had not directly intend. In the future, as we move into Humanity 2.0 mode, I believe that this theological argument, which is familiar from military justifications of 'collateral damage', will be increasingly invoked to justify the activities of 'humans' who seriously live up to having been created *in imago dei*.

Preparing for Life in Humanity 2.0 opens with a discussion of the philosophical foundations of Humanity 2.0, drawing attention to how recent changes in our understanding of the conduct of science and its social relations reflect implicit changes in human self-understanding more generally. Here three possible futures of 'being human' – ecological, biomedical and cybernetic – are sketched and ideologically interrelated. They recur in various forms throughout the book. Chapter 2 turns to Humanity 2.0's emerging political economy, which involves the redefinition of classical political and economic concepts, such as justice and productivity, with the aim of redefining the welfare state in a society where 'being human' is likely to involve a much wider range of embodiments than liberal societies have traditionally countenanced. Chapter 3 then looks at Humanity 2.0's 'anthropology', by which I mean the living conditions and aspirations available to this new being, which includes issues relating to the physical environment as well as standing as beings with both evolutionary and theological horizons. Chapter 4 is dedicated to Humanity 2.0's ethical horizons, centring on the normative sensibility of the 'moral entrepreneur', a natural risk-taker whose blurring of traditional intuitions of 'good' and 'evil' is bound to acquire greater significance and legitimacy in the future. The epilogue provides a revised general education curriculum that focuses on the West's changing attitudes to the brain, the organ that has historically provided the reliable concrete site for articulating the aspirations associated with Humanity 2.0.

DOI: 10.1057/9781137277077

I want to thank the wide range of people who, in rather different ways, over the past couple of years have encouraged me to develop the ideas and arguments contained in this book: Sarah Chan, Will Hunter, Campbell Jones, Veronika Lipinska, Luke Robert Mason, Willard McCarty, Nikolay Omelchenko, Ronald Porohyles, Hans Radder, Robert Ranisch, Chris Renwick, Gregory Sandstrom and Olga Stoliarova. Special thanks to Philippa Grand for including this book in the innovative Palgrave Pivot initiative. Finally, *Preparing for Life in Humanity 2.0* is dedicated to Rachel Armstrong, a true Renaissance woman who best exemplifies what it means today to live in the manner suggested in the title of this book.

DOI: 10.1057/9781137277077

1
Philosophy for Humanity 2.0

Abstract: *The nature of the human and the scientific have always been intertwined. In various sacred and secular guises, the unification of all forms of knowledge under the rubric of 'science' has been taken as the species prerogative of humanity. However, as our sense of species privilege has been called increasingly into question, so too has the very salience of 'humanity' and 'science' as general categories, My understanding of 'science' is indebted to the red thread that runs from Christian theology through the Scientific Revolution and Enlightenment to the Humboldtian revival of the university as the site for the synthesis of knowledge as the culmination of human self-development. However, the challenge facing such an ideal in the 21st century is that the predicate 'human' may be projected in three quite distinct ways, governed by what I call 'ecological', 'biomedical' and 'cybernetic' interests.*

Fuller, Steve. *Preparing for Life in Humanity 2.0.* Basingstoke: Palgrave Macmillan, 2013. DOI: 10.1057/9781137277077.

The nature of the human and the scientific have always been intertwined. Throughout the medieval and modern periods, in various sacred and secular guises, the unification of all forms of knowledge under the rubric of 'science' has been taken as the species prerogative of humanity. However, as our sense of species privilege has been called increasingly into question, so too has the very salience of 'humanity' and 'science' as general categories, let alone ones that might bear some essential relationship to each other. I begin this chapter by tracing this joint demystification to recent developments in the philosophy of science, which are neatly captured by what I call the 'Stanford School'. I then proceed on a more positive note to a conceptual framework for making sense of science as the art of being human. My understanding of 'science' is indebted to the red thread that runs from Christian theology through the Scientific Revolution and the Enlightenment to the Humboldtian revival of the university as the site for the synthesis of knowledge as the culmination of self-development. Especially salient here is science's epistemic capacity to manage modality (i.e. to determine the conditions under which possibilities can be actualised) and its political capacity to organise humanity into projects of universal concern. However, the challenge that faces such an ideal in the 21st century is that the predicate *human* may be projected in three quite distinct ways, governed by what I call *ecological*, *biomedical* and *cybernetic* interests. Which one of these future 'humanities' would claim today's humans as proper ancestors and could these futures co-habit the same world thus become two important questions that philosophy will need to address in the coming years.

1.1   Introduction: the road back from Stanford to a rehumanised science

There used to be a field called 'philosophy of science' that was focused on the set of epistemic practices that were most likely to realise humanity's deepest aspirations. It was concerned with defining and promoting the collective rationality of the species, otherwise known as 'progress'. The decline of this field can be traced to the influence of Thomas Kuhn (1970) on a generation of scholars, born around 1940, who started to become prominent in the history, philosophy and social studies of science in the late 1970s (Fuller 2000b: chap. 6). Within ten years, a critical mass of these scholars were assembled at Stanford University, centred on Ian

DOI: 10.1057/9781137277077

Hacking and Nancy Cartwright, and including such younger scholars as John Dupré and Peter Galison. Despite working in substantively different areas, they shared certain metatheoretic views: (a) anti-determinism and a more general scepticism about the reality of natural laws; (b) ontological pluralism as a pretext for methodological relativism and cross-disciplinary tolerance more generally; (c) a revival of interest in a localised sense of teleology and essentialism while renouncing more universalist versions of these doctrines; (d) a shift from physics to biology as the paradigmatic science and hence a shift in historiographical orientation from the Newton-to-Einstein to the Aristotle-to-Darwin trajectory; (e) a shift in empirical focus from the language of science to science's non-linguistic practices; (f) an aversion to embracing a normative perspective that is distinct from, let alone in conflict with, that of the scientific practitioners under investigation.

The Stanford School published a landmark volume (Galison and Stump 1996) that extended the reach of its line to fellow travellers of a more postmodernist, even posthumanist approach, as represented by, say, Donna Haraway (1991) and Bruno Latour (1993). The result has been the establishment of a diffuse but relatively stable consensus of bespoke thinking in science and technology studies (STS) that considers science in all its various social and material entanglements, without supposing that science requires an understanding that transcends and unifies its diverse array of practices. For these anti-generalists, as long as there are people called 'scientists' who refer to what they do as 'science', science continues to exist as an object of investigation. Indeed, the scientific agents need not even be people, provided that they are recognised by other recognised scientists as producers of reliable knowledge. In effect, rather than using sociology to flesh out a normative philosophical ideal, the Stanford School cedes to sociology conceptual ground that philosophers previously treated as their own.

My own philosophical position – known as *social epistemology* – stands in direct opposition to the Stanford School (Fuller 1988; Fuller 2000b: chap. 6; Fuller 2007b: chap. 2). However, the Stanford School is a useful foil for my argument because they recognise – if only to reject – the integral relationship between scientific unificationism, determinism, physics-mindedness and human-centredness. In what follows, I reassert all of these unfashionable positions but in a new key that acknowledges that the recent changes in the conduct and justification of science have coincided with a new sense of openness about what it means to be

DOI: 10.1057/9781137277077

'human'. I begin to explore this sense of openness in the next section, by considering how the rise of modern science has forced humanity to negotiate its species identity in terms of what is 'necessary' versus 'contingent' to our continued existence. In metaphysics this negotiation is the province of *modality*.

1.2   Science as humanity's means to manage modality

Grammatically speaking, modern science was born *concessive*, by which I mean that species of the subjunctive mood captured in English by 'despite' and 'although'. The original image conveyed by these words was one of modal conflict between overriding necessity and entrenched contingency, conceived either synchronically or diachronically: on the one hand, the resolution of the widest range of empirical phenomena in terms of the smallest number of formal principles; on the other, the unfolding of a preordained plan over the course of a difficult history. In short: Newton or Hegel. In both cases, necessity and contingency are engaged in a 'dialectical' relationship. While the 'necessary' might be cast in terms of Newton's laws or Hegel's world-historic spirit, anything that resisted, refracted, diverted or dragged such a dominating force would count as 'contingent'. What the Newtonian worldview defined as 'secondary qualities' or 'boundary conditions', the Hegelian one treated as 'historically myopic' or 'culturally relative'.

Both the synchronic and diachronic versions of this dialectic descended from divine teleology, where the end might be construed as either God's original plan (Newton) or its ultimate outworking (Hegel). However, modernity is marked by the removal of God as *primum mobile*, something that arguably Newton himself made possible once he defined inertia as a body's intrinsic capacity for motion (Blumenberg 1983). To be sure, de-deification has turned out to be a tricky move, since human confidence in science as a species-defining project is based on the biblical idea that we have been created 'in the image and likeness of God' (Fuller 2008b). In that case, the project of modern science may be understood as the gradual removal of *anthropomorphic* elements from an ineliminably *anthropocentric* conception of inquiry. By 'anthropocentric' I mean the assumption of reality's deep *intelligibility* – that is, reality's tractability to our intelligence. In other words, over successive generations, organised science has not only repaid the initial effort invested but

DOI: 10.1057/9781137277077

also issued in sufficient profit to inspire increased investment, resulting in a reconstitution of the life-world in the scientific image.

In the early modern era, the great Cartesian theologian Nicolas Malebranche provided a vivid metaphysical grounding for this sensibility by speaking of our 'vision in God', that is, our capacity to think some of God's own thoughts – specifically, those expressible in analytic geometry through which the motions of all bodies can be comprehensively grasped at once. Secular doctrines of *a priori* knowledge descend from this idea of an overlap in human and divine cognition. Moreover, the repeated empirical success of physicists' vaunted theoretical preference for the 'elegance' of mathematically simple forms cannot be dismissed as merely a tribal fetish. Those aesthetic intuitions are reasonably interpreted as somehow tapping into a feature of ultimate reality, the nature of which is not necessarily tied to our embodiment – but may be tied to some other aspect of our being.

Is the naturalist's explanation for science's success an improvement on the Cartesian one? After all, what exactly is the survival value of concentrating significant material and cultural resources on some hypothesised 'universe' that extends far beyond the sense of 'environment' that is of direct relevance to *Homo sapiens*? From a strictly Darwinian standpoint, a fetish that perhaps arose as a neutral by-product of a genetic survival strategy by a subset of the Eurasian population may only serve to undo the entire species in the long-term. Specifically, the mentality that originally enabled a few academics to enter 'The Mind of God' has been also responsible for nuclear physics and the massive existential challenges that have followed in its wake (Noble 1997; Fuller 2010a: chap. 1). In this respect, humanity's bloody-mindedness in the pursuit of science reflects a residual confidence in our godlike capacities, even after secularisation has discouraged us from connecting with its source.

Two senses of freedom are implied in this theological rendering of humanity's scientific impulse. On the one hand, we literally share God's spontaneous desire to understand everything as a unified rational whole, which drives us to see nature in terms of the laws by which the deity created. The clear intellectual and practical efficacy of this project stands in stark contrast to the risk to which it exposes our continued biological survival, since at least for the time being we are *not* God. This serves to bias our cost-accounting for science's technological presence in the world, whereby we tend to credit the good effects to the underlying science while blaming the bad effects on science's specific appliers or

DOI: 10.1057/9781137277077

users. On the other hand, our distinctiveness from God lies in the detail in which we seek a comprehensive understanding of reality, given our own unique status as creatures. This latter condition gives us a sphere of meaningful contingency, or 'room to manoeuvre' (*Spielraum*), to recall the phrase of the late 19th century German physiologist and probability theorist, Johannes von Kries, who greatly influenced Max Weber's thinking about what distinguishes humanity from the other lawfully governed entities in nature (Weber 1949: 164–88).

Humans are self-legislating, in that even in a world that is determined by overarching principles ('overdetermined', as it were), we have the power to resolve any remaining uncertainty in how these principles are instantiated. Indeed, Weber appeared to suggest that our humanity rests on the fact that we treat overdetermined situations as providing meaningful choices – that the way or style in which something is done in the short-term matters, even if in the long term it simply reinforces the fact that the thing had to be done. Vivid examples are provided by the extraordinary ethical debates that continue to surround birth and death – the simple biological fact that people come into and go out of existence. More striking than the range of opinion that these debates elicit is that they are had at all, since regardless of where one stands on abortion, contraception, euthanasia, suicide or – for that matter – murder, mortality is a certainty that befalls natural biological selves (Fuller 2006a: chap. 11).

This opening meditation on modality suggests that science relates to our humanity in a rather sophisticated way. We are dualistic beings, which in the first instance may be understood as our ongoing enactment of some version of spirit persevering in the face of recalcitrant matter. However else they differ, Newton and Hegel both belong to this moment. In that case, science's steadfast focus on necessity appears indifferent, if not cruel, with respect to the contingencies of particular human lives. But Max Weber – no doubt moved by the popular determinisms of his day (Marxism, psychoanalysis, energeticism) – recognised that these contingencies were precisely the means by which humans, even after conceding certain overriding tendencies in nature, express their distinctive identities. In that sense, for Weber, humanity is a variable by-product of the exigencies of the life-situations faced by our species. But there may also be collective by-products of those exigencies – and this is where what Karl Popper (1972) called 'objective knowledge', including science as a unique human achievement, belongs.

DOI: 10.1057/9781137277077

1.3   Science: a by-product of life that becomes its point and then imposes its own sense of justice

Karl Popper stood apart from most modern epistemologists and philosophers of science in refusing to identify knowledge claims – hypotheses and theories – with the formal expression of *beliefs*, which he took to be irreducibly subjective and more about self-affirmation than knowledge as such. Popper had an admirably literal, thing-like understanding of 'objective knowledge' as external entities one's contact with which is generative of systematic thought (Popper 1972). An attractive feature of this conception of knowledge is that its sense of objectivity is studiously neutral on the metaphysical debate between idealism and materialism (Fuller 1988: chap. 2). Thus, both Plato's forms and Popper's own example of the last library left standing after a nuclear holocaust would count as objective knowledge, in that each would enable thinking beings – whatever their provenance – to create a civilised world. Considering that, as we shall later see, humanity may proceed in at least two distinct directions in the future, Popper's view usefully refuses to tie science, understood as humanity's distinguishing feature, with our current biological makeup.

Popper's account of the origin of objective knowledge follows a line that in his youth had been advanced in the Vienna Circle as a transformation in the concept of 'economy'. It amounted to a definition of civilisation as the reversal of means and ends, once a means has achieved its end (Schlick 1974: 94–101). Thus, the practice of counting arose to solve problems related to survival but then, once those problems were solved (or at least managed), they became a project pursued for its own sake in the form of number theory. In the transition, the relevant sense of 'economy' shifted from producing ideas that minimise effort to a more focused concern for the minimal ideas needed to account for effort as such. Moreover, this long-term pursuit came to be seen as providing the basis for a still deeper economisation of effort that could be imparted in pedagogy. Here I mean the subsumption of the 'practical arts' in their broadest sense under 'science' – that is, the relation in which engineering, medicine, business and law currently stand to physics, biology, economics and the socio-political sciences, respectively, in the academic curriculum.

This general relationship is traceable to William Whewell's insistence that science requires properly trained 'scientists', not simply amateurs who happen to stumble upon epistemic breakthroughs. His philosophical legacy, the privileging of the 'context of justification' over the 'context of

DOI: 10.1057/9781137277077

discovery', came into its own in the second half of the 19th century. The distinction is ultimately about devaluing the contingent features of scientific achievement, so as to avoid a sense of 'path dependency' to science that would lose sight of its ultimate aim of universal knowledge (Fuller 2007b: chap. 2). In the short term Whewell's policies aimed to undercut the mechanics and inventors who over the previous century had flourished outside the clerically controlled academic networks. But in the long term his policies staved off – if not ensured – that scientific knowledge would be, at least in principle, made available to everyone, specifically those who had not undergone any idiosyncratic creative process or belonged to the right social network (Fuller 2000b: chap. 1). In effect, in trying to reassert the authority of Oxbridge in the face of interloping parvenus, Whewell struck a blow for *epistemic justice* (Fuller 2007a: 24–9).

In broadest terms, epistemic justice is about ensuring that individual inputs into the collective pursuit of knowledge are recognised in proportion to their relevance and credibility. Curiously, analytic philosophers frame this problem as one of *epistemic injustice*, namely, identifying and remedying presumptively clear cases in which the requisite sense of proportion has been violated (McConkey 2004). To be sure, such cases are easily conjured: medical research that studies only men but then pronounces on everyone; intelligence testing that fails to recognise its own cultural biases; psychological research that samples only students to draw inferences about all age groups. Research designs that systematically ignore significant segments of humanity and undermine science's aspiration to knowledge of universal scope and respect. Who could disagree? But the way forward is far from clear, which suggests that we need to get clear about what is meant by 'epistemic justice' before speaking of 'injustice'.

Consider the options available for a research design that claims to draw conclusions that are representative of *all* of humanity:

1   It must include a range of individuals who belong to the human sub-groups relevant to the research topic (Longino 2001).
2   It must include a range of the relevant perspectives on the research topic, regardless of who represents them (Kitcher 2001).
3   It must include people who can spontaneously adopt a sense of 'critical distance' on the research topic by virtue of having no necessary stake in whatever conclusions might be reached (Fuller 2006b: chap. 6).

DOI: 10.1057/9781137277077

Each option makes an implicit modal cut: that is, a line is drawn between what is necessary versus contingent in the properties of subjects in order to ensure that knowledge is produced of universal purchase. For example, (1) is committed to, say, flesh-and-blood women as representatives of a woman's point-of-view in a way that (2) is not. Yet, as long as women are allowed to represent themselves verbally (as opposed to by more direct physical means), it is entirely possible that their responses to research protocols will not deviate substantially from those of men (e.g. if they 'correct' their spontaneous phenomenology). In that case, perhaps some specially trained men (in Gender Studies?) would represent a woman's perspective better than actual women, just as a trained linguist might know an indigenous language better than an assimilated person of indigenous descent. Indeed, such a prospect might even be welcomed for preventing knowledge from being so closely tied to one's being in the world that it effectively becomes a source of *rent*, such that one cannot know a particular thing without being a particular person, in which case if one is not such a person, one needs to rent his or her services (Fuller 2010b).

For science to live up to its universalist aspirations, it must oppose regimes of 'information feudalism' generated by 'intellectual property' of all sorts, not least epistemic rent, as discussed above, which sometimes travels under the politically correct label of 'identity politics'. In all these cases, matters of epistemology would be converted to ones of ontology. Here the university stands as a bulwark against such conversion of knowledge to property with its Humboldtian mandate to incorporate the fruits of research into teaching, the overall lesson being that any knowledge originally acquired by one person can be, in principle, acquired by anyone (Fuller 2009). But the implications of this maxim for a more 'inclusive' science are not clear. Option (3) suggests that perhaps subjects should be allowed to make the modal cut for themselves by considering situations tangentially related to their self-understanding. They would thus need to both gauge the likelihood of a situation's relevance to their own lives and the difference it would make, were it relevant. The idea here would be to simulate a semi-detached but interested standpoint (Fuller 2007a: 110–14).

Here it is worth observing that models (1) and (2) – and quite possibly (3) – of epistemic justice sit uncomfortably with John Stuart Mill's classic argument in *On Liberty* for the free expression of thought as the optimal vehicle for collective truth-seeking. His particular formulation of this ideal assumed neither that people's beliefs are fixed by who they are (unlike 1) nor that certain beliefs deserved special representation in the polis (unlike 2). To be sure, Mill held that the expression of many different viewpoints contributed greatly to organised inquiry, and that special care had to be taken to ensure that minority positions were heard. However, he meant this activity to transpire entirely in the open, such that when decisions were taken, everyone knew where everyone else stood. In that way, people could draw their own conclusions about how others had reasoned, and on that basis perhaps adjust their own positions. Thus, in Mill's version of the open society, voting itself would be an open process, whose outcomes could be filed for future reference. A close analogue is the iterative character of the Delphi method developed by Nicholas Rescher (1998) and colleagues at the RAND Corporation in the 1950s to convert collective decision-making into a genuine learning experience that avoided the preemptive intellectual closure associated with 'groupthink'. As we shall now see, a groupthink-averse social epistemology is precisely what is needed for placing our future understanding of 'science' and 'humanity' in some reflective equilibrium.

1.4 Science's continual respecification of that projectible predicate 'human'

To speak of 'projectible predicates' is to recall that old warhorse of analytic epistemology, the so-called *grue paradox*. According to Nelson Goodman (1955), 'grue' is the property of being green before a given time and blue thereafter. This property enjoys just as much empirical support as the property of being green when hypothetically applied to all known emeralds. However, the sighting of some future emerald will distinguish whether emeralds had been always 'green' or 'grue', depending on whether it appears green or blue. For Goodman, this was a 'new riddle of induction' because unlike David Hume's original example of induction – how do we know that the sun will rise tomorrow just given our past experience – his problem suggests that our propensity to make inductive inferences is shaped not simply by our prior experience (which in the case of 'green' and 'grue' is the same) but by the language in which that experience has been cast, in which 'language' is taken to encode the structure of rational inference. Unfortunately Goodman drew a conservative conclusion from this situation, namely, that we are generally right to 'project' the more familiar predicate 'green' over 'grue' when making predictions about the colour of future emeralds. Why?

DOI: 10.1057/9781137277077

Well, because that predicate is more 'entrenched', which is a bit of jargon for the sub-philosophical stance of 'if it ain't broke, don't fix it'.

The prospect that a predicate like 'grue' might contribute to a more adequate account of all emeralds (both known and unknown) than 'green' is certainly familiar from the history of science. It trades on the idea that the periodic inability of our best theories to predict the future may rest on our failure to have understood the past all along. In short, we may have thought we lived in one sort of world, when in fact we have been always living in another one. (This characterises well the paradigm shifts from Ptolemy to Copernicus and from Newton to Einstein.) After all, the 'grue' and 'green' worlds have looked exactly the same until now. In this respect, Goodman showed that induction is about locating the actual world (where a prediction is made) within the set of possible worlds by proposing causal narratives that purport to connect past and future events, with 'green' and 'grue' constituting two alternative accounts vis-à-vis the past and future colour of emeralds. This is a profound point, especially for those 'scientific realists' who think that science aims at the ultimate representation of reality. Its full implications have yet to be explored – even now, more than half-a-century after Goodman's original formulation (cf. Stanford 2006 on the problem of 'unconceived alternatives' to the best scientific explanation available at a given time).

In particular, Goodman suggests how the 'paradigm shifts' that Kuhn (1970) identified with 'scientific revolutions' should be *expected* if we take the fallibility of our theories as *temporally symmetrical* – that is, that the outcome of any prediction of a future state has implications for what we believed about relevantly similar states in the past. In this respect, every substantial new discovery is always an invitation to do revisionist history. For, as science increases the breadth of our knowledge by revealing previously unknown phenomena, it also increases its depth by revising our understanding of previously known phenomena so as to incorporate them within the newly forged understanding. Thus, Newton did not simply add to Aristotle's project but superseded it altogether by showing that Aristotle had not fully grasped what he thought he had understood. Indeed, if Newton is to be believed, Aristotle literally did not know what he was talking about, since everything that he said that we still deem to be true could be said just as well – and better – by dispensing with his overarching metaphysical framework.

Science is usefully defined as the form of organised inquiry that is dedicated to reproducing Goodman's new riddle of induction on a regular

DOI: 10.1057/9781137277077

basis. Moreover, to recall the theological themes raised earlier in the chapter (which will also recur in Chapters 3 and 4), the achievement of a temporally symmetrical perspective on reality is one concrete sense in which scientists might inhabit the 'mind of God', insofar as one treats all points in time (not merely the ones through which one's human life transpires) as being of equal concern (Fuller 2010a: chap. 9). To be sure, speculatively conjured predicates like 'grue' rarely lead to successful predictions, so some skill must be involved to make them work so that they acquire the leverage for rethinking inquiry's prior history. Such skill is on display in first-rate scientific theorising. In this way, scientific revolutions metamorphose from Kuhn's realm of the unwanted and the unintended to Popper's (1981) positive vision of deliberately instigated 'permanent revolutions'.

One predicate whose projectibility will be increasingly queried vis-à-vis- developments in science is *human*. Already there is a small but interesting body of literature related to this topic, some arguing that science is confined to the human (Rescher 1999) and others that science exceeds the human (Humphreys 2004). But in both cases, 'human' is treated more like 'green' than 'grue'. In contrast, I propose that scientific inquiry is so bound up with what it means to be human that advances in science may render 'human' grue-like, such that David Chalmers' (1996) 'hard problem of consciousness' (which involves saving the phenomenology of human experience within a physicalistic worldview) is relegated to referring to accidental rather than essential features of our humanity.

Prima facie evidence for the grue-likeness of 'human' is provided by a recent on-line debate on 'the greatest technological advance of the 20th century' (*The Economist* 2010). The challengers were the computer and artificial fertiliser, the former extending certain human capacities for formal reasoning beyond their natural limits, the latter extending the quantity and quality of human lives to unprecedented levels. The debate was conducted largely in these terms, resulting 3:1 in favour of the computer. (I shall explore the political economy implications of this outcome in the next chapter.) A more philosophically salient indicator is the persistence of the mind-body problem: Those who would track the human through successive experiential states of our animal bodies in natural environments perennially meet resistance from those who hold that our humanity is only contingently tied to our embodiment and could, at least in principle, be realised – perhaps even more profoundly – in some other radically different medium, such as a computer. In the latter case, the human is associated primarily with the capacity to think, to reason and to reflect comprehensively

DOI: 10.1057/9781137277077

(i.e. 'consciousness' in the strongest sense of 'self-consciousness'), where the primary terms of reference do not depend on a particular material instantiation. Increasingly prominent as a battleground for these conflicting intuitions is the concept of 'dignity', perhaps the quality associated with our humanity that is most directly tied to our bodies (Cochrane 2010).

Thus, put starkly, the relevant 'green' versus 'grue' alternatives consist in whether the future of the human should be projected through (1) the reproductive patterns of *Homo sapiens* and our evolutionary successors, even if natural selection turns out not to favour what we currently regard as our distinctive mental powers, or (2) whatever physical means is required to preserve and more fully realise those distinctive mental powers, even if they involve shifting from a carbon to a silicon base altogether. In terms of these alternatives, the philosopher of animal liberation Peter Singer (1999) would represent the first projection of the human and the transhumanist guru Ray Kurzweil (1999) the second. Key indicators here are, on the one hand, Singer's 'levelling' of the threshold of moral relevance from a unique human power such as rational agency to pain avoidance, a state that humans share with all animals in possession of a nervous system and, on the other, Kurzweil's view of death as representing more a remediable technological shortcoming than a natural state of the human condition (Fuller 2011: chap. 2).

The middle ground between these two extremes is occupied by the programme of *eugenics*, whose advocates from Francis Galton to Julian Huxley have aimed to preserve and extend humanity's distinctive mental traits, while staying within the parameters of our biological heritage. The idea was that strategic interventions into both our reproductive capacity and nature's selection pressures would enable us to direct the course of evolution. The idea is *prima facie* plausible if we truly believe that *Homo sapiens* has had more impact on the planet in recent aeons than any other species. Over the past decade, science-policy initiatives on both sides of the Atlantic to foster the 'convergence' of nano- and biotechnology research have given eugenics an ideological makeover, mainly by shifting the locus of normative concern from the population to the individual, as befits our 'bioliberal' age (Fuller 2006a: chap. 1). Thus, 'negative eugenics' applied to a population is now rebranded as 'therapeutic' when applied to the individual, and similarly 'positive eugenics' as 'enhancing' (Fuller 2011: chap. 3). Although much ink has been spilled about the dehumanising and irreligious character of eugenics research and policy, in fact many of its most distinguished contributors were, broadly speaking,

DOI: 10.1057/9781137277077

non-conformist Christians. Indeed, taken at face value (and not in terms of the political excesses it spawned), eugenics is most naturally understood as extending the idea of humanity's Biblical stewardship of the earth from agriculture to human culture. It is not by accident that a notable early impact of eugenics-based thinking was 'puericulture', the science (or, perhaps better, technology) of 'raising children' in the literal sense of raising plants or animals (Pichot 2009).

Of these three projections of 'human' – Singer's posthumanist evolutionary mutation, Kurzweil's transhumanist android, and the eugenicist's middle ground – we may ask a question inspired by George Sarton, the founder of professional science history, to update Auguste Comte's original positivist vision of the history of science as the narrative through which humanity's collective self-realisation is told (Sarton 1924; cf. Fuller 2010a: chap. 2): *From which (if any) of these visions are likely to come beings who would be recognised by the historic exemplars of our humanity as their descendants?* The intuition guiding the question is clear and reasonable: If we claim that, say, Newton or Goethe exemplifies our sense of humanity, then we should imagine that, given the opportunity, he would return the compliment and accept us as his intellectual offspring. Such thought experiments in mutual recognition may be seen as a more analytic and normative version of the hermeneutical 'fusion of horizons' that Gadamer held to be a precondition of historical understanding (cf. Fuller 2008a). These exercises bear on the projectibility of 'human' because they force us to imagine ourselves as, say, Newton or Goethe passing judgement on whether we have adequately captured – if not amplified – the spirit (aka essence) of their original projects. Insofar as he would recoil from any of the three projected versions of 'Humanity 2.0', we are then faced with the following choice: (a) to dissociate ourselves from his legacy (and perhaps try to locate some intellectually more congenial 'humanist' ancestor); (b) to admit that there are now two species of 'humanity' that nevertheless share a common ancestor; (c) to abandon 'humanity' altogether as the conceptual banner under which our collective project travels.

If the reader is disoriented by this array of choices, the reason may be that I take seriously the prospect that, if 'humanity' is identified with our species-distinctive mental traits that are only imperfectly realised in our current biological form, our successor being may not look or feel very much like us at all – especially if Ray Kurzweil gets his way. For example, traditional virtues such as prudence and compassion have

DOI: 10.1057/9781137277077

been compelling for beings with roughly our bodies whose experiences are confined to familiar social environments. But the meaning of these virtues are already being stretched as people who are still grounded in the bodies of their birth move in increasingly heterogeneous social environments, courtesy of air travel, mass media and cyberspace. Now, what happens once these people loosen their self-identification with their biological bodies, as in the phenomenon that Sherry Turkle (1984) originally called 'second self', which has now been normalised as 'Second Life' avatars? If the value placed on an activity is measured in terms of the time spent on it, then many people appear to be in the process of intellectually, if not physically, evacuating their biological bodies.

A general philosophy of science curriculum is required for this emerging Humanity 2.0. Its aim would be to provide an intellectual framework to ensure that the three projections of the human outlined above are made the subject of educated judgements and not simply allowed to track the market forces that currently guide their development – specifically, those driven by *ecological interests* (for Singer's posthumanism), *biomedical interests* (for the new eugenics) and *cybernetic interests* (for Kurzweil's transhumanism). Two hundred years ago, the German idealists first gave philosophy a clear disciplinary mission – namely, to provide the metatheory of the university by articulating what qualifies the various bodies of disciplined knowledge as 'science'. The practical side of this task was executed by an integrated course of study in which the student would learn enough about each of the disciplines to arrive at a personal synthesis, a 'science of oneself', as it were. Here philosophy offered principles for unifying knowledge in the service of life. In this context, Goethe was seen as superior to Newton as a human being for having engaged this task more thoroughly and creatively. A 'general philosophy of science' for our own time would reforge the idealist link between the unity of the self and the unity of science – but in a new key. Instead of the familiar post-Kantian task of integrating the sciences of 'nature' and 'spirit' (i.e. *Natur-* and *Geistes- wissenschaften*), we now face the more daunting challenge of designing a scheme that draws together the increasingly marked ecological, biomedical and cybernetic interests that are charting the course of Humanity 2.0. In the epilogue I address this challenge by proposing a brain-centred model for university-level 'general education' in the 21st century.

DOI: 10.1057/9781137277077

2
Political Economy for Humanity 2.0

Abstract: *Humanity is historically poised to renegotiate its sense of collective identity, as reflected in five emergent trends: (a) the prominence of digital technology in shaping everyday life and human self-understanding; (b) the advances (both promised and realised) in biotechnology that aim to extend the human condition, perhaps even into a 'trans-' or 'post-' human phase; (c) a growing sense of ecological consciousness (promoted by a sense of impending global catastrophe); (d) a growing awareness of the biological similarity between humans and other animals; (e) an increasing sense of human affection and sympathy migrating to animals and even androids, during a period when national health budgets are stretched perhaps to an unprecedented extent. After exploring the humanistic implications of the computer's overriding prominence as an innovation in the 20th century, I draw out the implications for welfare policy more specifically, concluding with a thought experiment concerning health policy that challenges our intuitions about who or what is to be included as human in tomorrow's 'liberal' society.*

Fuller, Steve. *Preparing for Life in Humanity 2.0.*
Basingstoke: Palgrave Macmillan, 2013.
DOI: 10.1057/9781137277077.

DOI: 10.1057/9781137277077

Humanity is historically poised to renegotiate its sense of collective identity, as reflected in emergent trends in the political and economic discourse. Five sources for these trends are addressed in this chapter: (a) the prominence of digital technology in shaping everyday life and human self-understanding; (b) the advances (both promised and realised) in biotechnology that aim to extend the human condition, perhaps even into a phase that might be called 'trans-' or 'post-' human; (c) a growing sense of ecological consciousness (much of it promoted by a sense of impending global catastrophe); (d) a growing awareness of the biological similarity between humans and other animals, reviving doubts about strictly naturalistic criteria for demarcating the 'human'; (e) an increasing sense of human affection and sympathy migrating to animals and even androids, during a period when national health budgets are stretched perhaps to an unprecedented extent. The chapter is structured in two parts. The first part follows the humanistic implications of the claim that the computer was the innovation that most changed the human condition in the 20th century. The overriding significance of the computer provides a gateway to our emerging sense of 'Humanity 2.0'. The second part focuses on the implications of Humanity 2.0 for welfare policy, concluding with a thought experiment concerning health policy. Here the basic point is that the ontological framework for conceptualising the just liberal society is subtly shifting from *the potential* to *the virtual* as the normative benchmark of our humanity.

2.1   What the computer says about who we think we are: a portal to Humanity 2.0

You can tell a lot about the sort of creature we think we are by the value we place on the things we make. In October 2010, a widely watched on-line debate was staged on the most important technological innovation of the 20th century (*Economist* 2010). The challengers were the *digital computer* and the *artificial fertiliser*. (30–60 years earlier, *nuclear power* would have been a contender and possibly the winner.) Perhaps unsurprisingly, the computer won by a margin of 3 to 1. But why is this not surprising? After all, the artificial fertiliser is arguably the invention most responsible for a fourfold growth in the world's population over the past century, as well as reducing the proportion of those suffering from malnutrition by at least two-thirds. It would be difficult to think of another product of human ingenuity that has had such deep and lasting

DOI: 10.1057/9781137277077

benefits for so many people. And even if it is true that in absolute terms there are more people living in poverty now than the entire population of the earth in 1900, the success of artificial fertilisers has kept alive the dream that all poverty is ultimately eradicable.

Yet, the artificial fertiliser was trumped by the computer – even though the computer's development has tracked, and in some cases amplified, global class divisions. Indeed, it is becoming increasingly common to speak of 'knows' and 'know-nots' in the way one spoke of 'haves' and 'have-nots' 50 years ago (Castells 2009). Here it is worth contrasting, on the one hand, the Protestant literacy drive accompanying the development of the printing press that over time served to reduce the power asymmetry between the producers and consumers of writing with, on the other hand, the tendency – most successfully promoted by that great computer illiterate Steve Jobs – to discourage digital denizens from having to learn programming code to get what they want from their gadgets (Appleyard 2011: chap. 6). Thus, the US media theorist Douglas Rushkoff (2010) has felt compelled to compose a digital call-to-arms entitled 'Program or Be Programmed'. Its shock value trades on the absence of an earlier manifesto called, say, 'Write or Be Written About' – even though the epistemic authority of empirical social science over its subject matter arguably depends on just this asymmetry. Put more pointedly: It is easy to imagine a counterfactual Steve Jobs associated with the 16th century print revolution who would have promoted a 'path of least resistance' consumer-oriented literacy policy akin to bluffing one's way through a foreign language by capitalising on cognate forms without ever properly mastering its grammar and semantics. Such a feat would require skills comparable to navigating Apple's sleek user-friendly interfaces. But would this counterfactual Jobs have been as celebrated as the actual one?

Over the ten days of debate at the *Economist* website it became clear that the computer was bound to triumph because, for better or worse, we identify more strongly with the *extension* than the *conservation* of the human condition. Whether we categorise this extension as 'culture', 'technology' or, in Richard Dawkins' (1982) case, something still cast in Darwinian terms, the 'extended phenotype', it suggests that we are not fully human until or unless our biological bodies are somehow enhanced, if not outright transcended. The computer captures that desire in a twofold sense: It both provides a model for how to think of ourselves in such an enhanced state and the environment in which to realise it. UK media theorist David Berry (2011) has explored the implications of this development in terms of

DOI: 10.1057/9781137277077

such computer-based technologies as iPhones and iPads that increasingly constitute the human life-world. Bluntly put, the more time people spend interacting with high-tech gadgets, the more grounds there are for claiming that what the previous generation called 'virtual reality' is becoming the actual reality in which people define themselves.

Thus, it is not surprising that an invention that 'merely' keeps alive our normal biological bodies – the artificial fertiliser – should be ranked decidedly lower than the computer in terms of importance. Indeed, the artificial fertiliser may have contributed to a compounding of humanity's problems in the 20th century by enabling a relatively superficial level of survival for the species without adequate planning for their long-term flourishing. From that perspective, the persistence of poverty noted above may be explained not in terms of a lack of resolve to follow through on one's good intentions but an original failure (sin?) to recognise the incapacity of human nature to live up to its own ideals. Drawn to its logical conclusion, as done by Darwin and Spencer, and more recently Peter Singer (1999) and Steven Pinker (2002), this line of reasoning sees the sort of 'idealism' associated with the desire to expand and prolong the ranks of *Homo sapiens* as a narcissistic denial of our fundamental animal nature that only promises to keep us forever in a state of tantalised torment. In that case, would it not be better to go with the grain of nature and simply give up any hope that we might turn the planet into what the Jesuit scientist Pierre Teilhard de Chardin (1961) called a 'hominised substance'?

There is evidence that individuals are already doing that. Nearly a half-century ago the economist Thomas Schelling (1968) argued that you can tell the value that people place on their own lives by the amount they are willing to pay for securing it. The argument was originally taken to show that large defence budgets worldwide meant that people placed a sufficiently high value on their lives to render a Third World War unlikely. But considered *today*, whether 'security' is defined in terms of healthcare, life insurance, development aid or military budgets, one would be left with an open verdict on the exact value that people place on the indefinite maintenance of the bodies of their birth. Despite the lip service paid to the idea of a long and safe life, there is little neuropsychological evidence for the spontaneous inclination to make the necessary savings along the way, absent coercion, say, in the form of state taxes (Aharon and Bourgeois-Gironde 2011).

So, if we identify people's preferences with what they do rather than what they say, it would seem that beyond a certain point, people prefer

DOI: 10.1057/9781137277077

to forgo physical security in favour of the freedom (and risk) to explore alternative possible modes of existence. In this context, the computer is the paradigmatic technological portal, allowing for varying degrees of 'transhumanisation': from, say, the somewhat idealised web-based self-representations that are now commonplace, through a 'Second Life' avatar that enables one to live a fantasy-self alongside the actual self, to Ray Kurzweil's (2005) vision of our uploaded consciousness in machines that can function indefinitely. There is evidence that even the developing world has bought into a scaled down version of this line of thought. For example, the number of mobile phones has outstripped that of clean toilets worldwide, even in India, an emerging economic powerhouse – and the penetration of mobile phones into the population of some Third World countries outstrips that of the United States (Giridharadas 2010).

Declaring these 'impoverished' mobile phone users as suffering from 'false consciousness' or some other psycho-political pathological sounds like the death rattle of the Enlightenment recipe for achieving universal human progress. Indeed, we need to take seriously that people may order their existential needs differently from, say, those enshrined in Abraham Maslow's (1954) self-actualisation theory: to wit, that the capacity to communicate with others is valued more highly than the security of one's own material conditions. Interestingly, this prospect was foreseen over 30 years by public choice economists who argued that the desirability of meeting supposedly 'fundamental' life needs, such as secure food, water and shelter, maybe be offset by the lower immediate costs and higher immediate benefits involved in meeting 'higher-order' needs, as allowed nowadays by being plugged into social media (cf. McKenzie and Tullock 1981: chap. 20). Faced with such a choice, people may well decide to put themselves at long-term material risk in a manner perhaps not so different from those who would debase their bodies for an appropriate spiritual experience.

The normative significance so attached to the computer puts paid to one popular projection of a 'transhuman' future, whereby we become able to slow or even reverse the ageing process through gene-based interventions (De Grey 2007). This prospect has been a source of great individual hope but also great collective fear – especially from the standpoint of welfare provision and ecological sustainability. Arguably much of the urgency surrounding both the converging technologies and the climate change agendas today is fuelled by the image of an increasing number of high-producing, high-consuming humans of indefinite longevity in the not too distant future (Fuller 2011: chap. 3). But it is by

DOI: 10.1057/9781137277077

no means clear that this image is widely held beyond the precincts of its upwardly mobile 'middle youth' proponents who, say, populate the audiences at TED (Technology, Entertainment and Design) lectures that have been emanating from California's Silicon Valley since 1984. Those who are younger and older, as well as poorer, than TED's target demographic might simply find a computer-based future more appealing than maintaining one's biological body indefinitely. Indeed, there is no clear reason why people would want to live a long time, once it becomes socially acceptable (aka legally recognised) to 'live fast, die young'. We are already seeing a steady neglect – perhaps even abandonment – of the embodied human by those who spend most of their lives in front of a computer screen. This goes beyond the fact that obesity and heart disease are back on the rise in the First World (as well as occurring for the first time in newly developing countries).

At the same time, there is a growing underground trade in drugs originally designed to repair mental and physical deficiencies, but now retooled and remarketed to 'enhance' normal performance. Typically travelling under the rubric of 'open source' or 'do-it-yourself' biology, this development's democratic self-understanding masks the substantial risks often willingly undertaken by self-experimenting individuals (Hope 2008). To be sure, over the past half-century, there have been precedents in the development of amphetamines (Rasmussen 2008) and LSD (Langlitz 2010). Moreover, the adventurous end of bioethics has presented reasonable arguments that current 'Institutional Review Board' (IRB) constraints on human-based scientific experiments should be augmented, if not outright superseded, by a legally enforceable human 'right' or 'duty' to involvement in such experiments (Chan et al. 2011). (IRBs are university committees that are empowered to prohibit research that might entail morally inappropriate treatment of subjects.) The fundamental intuition here is that true individual liberty entails the personal assumption of risk: If people are 'free' to spend most of their disposable income on lotteries that they have little chance of winning, why cannot they be 'free' to subject themselves to research that typically has a better – though still often less than 50 per cent – chance of improving their lives? One might add to this case a sense of personal responsibility for the welfare of society as a whole: After all, even a treatment that fails to improve the lives of those who undergo it will have set a negative example to be avoided by others in the future. In contrast, one's failure to pick the winning lottery number is designed precisely *not* to allow for such a collective learning experience.

DOI: 10.1057/9781137277077

At a deeper level, of course, there is nothing new about treating our biological bodies as living laboratories. Any mass change in a population's dietary regime has had long-term psychotropic effects, typically marked by a shift in the default expectations of normal behaviour (Smail 2008). Of special significance is the shift that occurred in Europe over the 18th century, whereby a diet consisting of constant low levels of alcohol consumption (partly as a water purification strategy) was replaced by one with constant low levels of caffeine consumption (via coffee and tea), alongside more concentrated doses of alcohol consumption (via spirits). This period corresponds to the Enlightenment, when radical ideas – many of which had been in the air for at least a century – began to acquire a vividness that inspired organised action, not least revolutionary overturns of ancient regimes. And even though many of these efforts turned out to be abortive, a step change had taken place in the Western collective psyche. Life's baseline pulse was quickened, such that by the 19th century the similarity of the present to the past came to be seen less as reassuring than stultifying, and products of the imagination ranging from scientific theories to economic policies were granted greater license for changing the world.

At the same time, our capacity to suppress and mask the novel forms of pain and suffering that accompanied these transformations also increased. The mass consumption of aspirin and plastic surgery in the wake of the First World War perhaps marked the tipping point into today's mindset, which is averse to any routine form of pain or suffering – indeed, to such an extent that it is now a commonly accepted standard of moral relevance, with those who inflict or suffer pain indefinitely regarded presumptively as evil or perverse (Singer 1993). But here it is worth recalling that both aspirin and plastic surgery were originally designed to minimise, if not over time erase, memories of military experience so that soldiers who had survived the First World War could function as confident citizens in peacetime. Not surprisingly, then, these innovations were quickly marketed in the 1920s as means to increase one's positive outlook in an increasingly complex and competitive world (Chatterjee 2007). In other words, originally the pain threshold was treated as a Nietzschean challenge to be overcome in the spirit of 'what doesn't kill me makes me stronger'. But as the 20th century wore on, no doubt influenced by the anxieties generated by the Second World War and the Cold War (and extending into the ambient 'War on Terror' of our own times), the pain threshold came to be regarded in a more Schopenhauerian fashion as

DOI: 10.1057/9781137277077

something to be avoided at all cost, even if that means suicide, say, once the doses no longer have the desired effect.

While many have pathologised humanity's increasing willingness to treat the body as a biochemical testing ground, the phenomenon may also be seen as part of a broader cultural transition – a shift from what the transhumanist philosopher Max More (2005) has called a *precautionary* to a *proactionary* worldview: the former risk-averse, the latter risk-seeking. The ascent of the proactionary perspective points to an impending and profound division in lifestyles between those who would ideally live in perpetuity and those who would wish to make their mark as soon and sharply as possible regardless of longevity. The former, one might say, are 'second-order' proactionaries who do not take the risks themselves but learn from those taken by the latter, the 'first-order' proactionaries. The challenge then will be to reap the most social benefit from those latter 'shooting stars' so that their lives are not merely self-consuming. In principle, this should not be too difficult. Since such people, even if they are not participating in cutting edge experiments, will be leaving tracks throughout cyberspace, each one's 'death' may be seen as an absorption into virtual reality (what the 'Second Generation' of *Star Trek* called a 'borg'), a passage from one's carbon-based bodies to an immortal silicon existence. Backing this specific rite of passage is long-standing psychological evidence that genius in various fields tends to be exhibited at particular ages, after which the geniuses fall into often pathological decline, at least partly due to their felt inability to maintain that level of performance throughout their lives. While anecdotally most closely associated with mathematics and poetry, those fields are distinctive only in the relative youth of the onset of such feelings, but the feelings themselves are widespread across all fields (Simonton 1984).

A precedent for such self-affirming planned obsolescence is the euthanasia 'departure lounges' in Richard Fleischer's 1973 dystopic film, *Soylent Green*. In this case, those already with significant achievement to their credit need not bear the burden of surpassing it, once they are provided with an opportunity to witness their digital immortalisation, whereby they are captured in their prime forever, without the need to record years of decline and degradation. The price, of course, is that they agree to die now. In that case, tomorrow's 'welfare safety net' may be to do mainly with protecting one's posthumous reputation by securing a place in humanity's collective narrative. This 'politics of recognition' would mark a turn away from the Hegel-inspired version popularised by

DOI: 10.1057/9781137277077

Marx as an explicit political demand made on behalf of some subaltern social category, which is supposed to be met by one's contemporaries, say, through 'affirmative action' or 'positive discrimination' legislation. Instead it would mark a return to the original Greek preoccupation with posthumous fame as the ultimate form of respect, but now understood as repeatable invocation in anecdotes or gossip or, increasingly, digital hits (Fuller 2006a: chap. 9). Taken to the limit, one might envisage here a revaluation of what Hegel originally derided as 'monumentalism', the first moment in aesthetic history, exemplified by the Pyramids of Egypt, those perpetual memorials for great souls that succeeded only in establishing death cults. But now we have 'BioArt', the radical end of which envisages that a significant aspect of a person's identity (organic material, genetic code or digitised memory) might be embedded 'informatively' into an artefact that outlasts the original person (Mitchell 2010). There is already precedent in the idea of 'living architecture', whereby organic (typically plant) material is inserted into buildings to enable them to respond more flexibly to environmental changes (Armstrong 2012). Perhaps in the future a building will be named for someone who gave not merely money but some of her very being to its construction and maintenance.

The idea that the human condition might transition from one of biology to technology revives the theological impulse that drove the Scientific Revolution of 17th-century Europe, namely, that we understand the totality of life as an artefact that bears the Creator's design (Fuller 2010a), a secular version of which continued to inform Kant's *Critique of Judgement*, which systematically pursues the analogy of 'purpose' in art and nature. But even if we stay within the history of biology proper, the idea of humanity as literally a 'work in progress' reigns supreme. Once the discipline entered its current scientific phase by abandoning a typological approach to species (understood as either Platonic archetypes or Aristotelian natural kinds) in favour of a populational approach (common to both Darwin and Mendel), the default position in the normative status of the individual, human or otherwise, vis-à-vis the collective also shifted. The individual came to be seen 'instrumentally' in one of two senses: either as a means to improve the species or, more simply, a vehicle for reproducing the species (Fuller 2011: chaps. 4–5).

To be sure, 'species' does not have quite the same meaning in the two cases: the former presumes a clear *telos* that the latter does not. We now think about this difference in terms of Lamarck versus Darwin but the original population theorists – Marquis de Condorcet and Thomas

DOI: 10.1057/9781137277077

Malthus – adopted these two positions, respectively. Thus, Condorcet valued unlimited population growth as an extension of the 'two heads are better than one' principle, whereas Malthus regarded population as the stage on which God displays his mastery over nature, which humans are capable of grasping in statistical terms (Fuller 2006a: chap. 13). The difference here is largely theological: As heretical Catholics, Lamarck and Condorcet share the Pelagian idea that humans can voluntarily achieve salvation, perhaps even by responding solicitously to perceived divine cues. In contrast, Darwin and Malthus presuppose the more Calvinist idea that individual survival is entirely in God's hands (aka natural selection), in which case our behaviour counts merely as data for the deity, not acts of persuasion that might turn the divine mind in humanity's favour (hence the 'blindness' of natural selection). To be sure, in *both* cases, the individual is ultimately sacrificed for some imagined end that is presumed to be of collective benefit, however that collective is defined. Thus, a revival of *theodicy*, the theological discipline concerned with God's sense of justice, is likely to be an integral part of the worldview of 'Humanity 2.0', as will become apparent in the rest of this book.

2.2 Redefining welfare for Humanity 2.0: from the potential to the virtual human

To privilege the computer as a medium of human self-expression is to challenge our well-being on at least two levels, which might be called *intensive* and *extensive*, to be explained below. Conceptually speaking, 'intensive' magnitudes presume a limit or end in terms of which one may be nearer or farther, whereas 'extensive' magnitudes exist along a dimension that may be increased indefinitely. (The underlying practical intuitions concern, respectively, measuring and counting.) But with regard to intuitions about human well-being, the natural integrity of the human body is not taken as a stable baseline intuition. In this context, it is useful to speak of an emerging worldview of 'Humanity 2.0', where 'Humanity 1.0' consists of those autonomous but sociable individuals that we normally imagine ourselves to be. Humans 1.0 are the beings that our laws have traditionally been designed to empower and protect. This aspiration was finally given global recognition after the Second World War in the United Nations Declaration of Human Rights (1948). Of course, as demonstrated by the subsequent diplomatic condemnations, trade embargos

DOI: 10.1057/9781137277077

and occasional wars, the Declaration has not been sufficient to secure the well-being of Humanity 1.0. Nevertheless, it has provided a normative standard against which regimes and policies have been legitimately judged on the world stage.

One might imagine that any ideas of 'Humanity 2.0' would build upon the principles enshrined as 'Humanity 1.0'. However, this is not the case – not only because most of the world's population still lives without the material prerequisites for human dignity outlined in the Declaration and – not only because many of those people, as we saw in the last section, seem themselves to prefer having easy access to information and communication than secure food, water and shelter. Most tellingly, the concept of dignity itself is under fire as unfit for purpose as a defining characteristic of human well-being, at least if the editors of the journal *Bioethics* are to be believed (Schüklenk and Pacholczyk 2010). For them – and they are hardly alone (see also Pinker 2008) – 'dignity' offers little more than a euphemistic invitation to think of humanity in relatively static natural law terms that sharply distinguish, say, 'therapy' and 'enhancement' as goals of biotechnological interventions, whereby the former refers to restoring someone to their original ('natural') state and the latter to some hypothesised improved ('artificial') state (Fukuyama 2003). Enforcement of this distinction lay behind George W. Bush's withholding of US federal funding from stem-cell research (Briggle 2010).

More specifically, Humanity 2.0 challenges our sense of the human from an *intensive* standpoint, such that what distinguishes the 'human' from the 'non-human' is increasingly subject to degrees and variation. This is happening in the context of a neo-liberal political economy, in which what is 'normal' (either statistically or normatively) is subject to market forces. Thus, a drug that we might now consider 'brain boosting' because it enables performance that exceeds the norm may itself set the norm in the future, simply by virtue of increase uptake or even aspiration to uptake. This prospect is governed by an ideology of 'able-ism', that is, that we end up being 'always already disabled' as the norm of competent performance drifts upward (Wolbring 2006). Health will have become a 'positional good', whereby our sense of well-being is tied directly to our comparative advantage vis-à-vis others (Hirsch 1976). Indeed, the most recent edition of the American Psychiatric Association's *Diagnostic and Statistical Manual of Mental Disorders* (DSM-5), due for publication in 2013, appears designed to promote this fluidity of normal human performance, much to the consternation of Allen Frances, the chief editor

DOI: 10.1057/9781137277077

of the previous edition (Greenberg 2011). In particular, an increasing number of mental disorders are characterised – *à la* Freud *vis-à-vis* neuroses – in terms of a sliding scale or spectrum, as well as a larger range of 'risky' environments that might trigger a mental disorder. In the latter case, as we shall see below, the emerging science of 'epigenetics' purports to provide a biological basis.

Behind these moves to relativise quite radically the distinction between the 'normal' and the 'pathological' is a fundamental ambivalence on the part of humans towards the bodies of their birth. After all, when compared with other animals, we take a long time to reach adulthood. This bare fact has led philosophers down through the ages to muse that we are by nature premature beings who must go beyond our native biology to complete our existence. The 20th century witnessed the emergence of a school of thought – originally called 'philosophical anthropology' and later 'general systems theory' – dedicated to the systematic exploration of what follows from our biological incompleteness (Bertalanffy 1950). This project has been arguably operationalised in recent years under the rubric of *epigenetics*, an empirical investigation into the finalisation of the exact genetic makeup of individuals, which usually occurs in infancy and early childhood. The last major thinker to have taken this idea seriously was Sigmund Freud, whose claims for an Oedipal and Electra Complex was the source of ridicule by much of the scientific establishment for most of the 20th century. To be sure, Freud unfamiliar with the molecular basis of genetics, held the Lamarckian view that traumatic memories as such could somehow shape the genome. Nevertheless, he managed to elude the spell that the Weismann Barrier – that somatic changes can never *ipso facto* result in genetic changes – had cast on 20th century evolutionary theory, from which we are emerging only now (*Economist* 2006).

But even with the Weismann Barrier in place, it was thought that we might beat the genetic odds stacked against us by mastering the laws of heredity. This view was of course associated with the eugenics programme of Francis Galton, whose cousin Charles Darwin refused to endorse. Nevertheless, such 'Genetically Modified Darwinism', so to speak, survived as what may now be seen as the underground prehistory of transhumanism. A trajectory that includes various attempts – say, from US developmental psychologist James Mark Baldwin to the UK animal geneticist Conrad Waddington – to simulate Lamarckian evolution by strictly Darwinian means (Dickens 2000). Their broad common goal was

DOI: 10.1057/9781137277077

to explain the apparent inheritance of acquired traits as a macro-level consequence of selection pressures on a population by proposing that those individuals already capable of expressing the requisite traits in a changed environment are reproductively advantaged. Thus, over time economic classes could morph into biological races that amount to a caste system. Of course, the meaning of 'inheritance of acquired traits' has become more fluid with recent advances in gene therapy and other forms of biotechnology that permit strategic intervention (what theologians might regard, perhaps with some consternation, as 'intelligent design') at both ante- and post-natal stages. It is in this context that one nowadays often speaks of epigenetics as reviving the early modern idea of 'epigenesis' (Shenk 2010). In its original 18th century context, epigenesis was one side of a dispute about how to interpret microscope-based experiments concerning an organism's pattern of development. The other side was preformation. At stake, we might now say, was whether development was 'matter-led' (*epigenesis*) or 'form-led' (*preformation*). In the former case, the maturing organism is open to multiple paths of development, which in the end is determined by environmental input. In the latter case, the organism has a predetermined path of development that ultimately overcomes whatever interference the environment provides. In the language of general systems theory, classical epigenesis was committed to *plurifinality*, preformation to *equifinality*. I have referred to *underdetermination* versus *overdetermination* to capture this difference in casting the narrative of life as a general feature of historiography (Fuller 2008a).

The two great 19th century evolutionists, Lamarck and Darwin, both more concerned with the natural history of entire species than the life trajectory of particular individuals, may be seen as having held opposing combinations of these two views. Lamarck was an epigenesist about individual lives but a preformationist about natural history as a whole: He believed that individuals could improve upon their inheritance in ways that brought their offspring closer to some ideal state of being. In contrast, Darwin was normally read as a preformationist about individual lives but an epigenesist about the overall course of natural history: For him, the largely genetically fixed nature of organisms discouraged any hope for indefinite survival, let alone improvement, against an environment possessing no concerns of its own, let alone those of particular organisms.

The emerging science of epigenetics renders the workings of epigenesis more transparent and hence more controllable, which arguably means

DOI: 10.1057/9781137277077

that it can simulate the directionality implied in the preformationist perspective. Unlike the 20th century-style 'Genetically Modified Darwinism' discussed above, epigenesis-based policies favouring transhumanism need not be limited to realising individual genetic potential by matching the right genomes to the right general environments. Rather, they can draw on the finding that genomic expression as such requires exposure to the environment, especially via the chemical process of 'DNA methylation', which has been experimentally induced in animals to switch specific genes on and off (Borghol et al. 2011). By undergoing this specific process, so say today's epigenesists, organisms come to possess determinate traits, which they then maintain in the face of subsequent environmental changes and their offspring find easier to express.

At first glance, epigenetics appears to shift evolution's horizons away from Darwin's back to Lamarck's, except that Lamarck portrayed animals as deliberately changing their genetic makeup through willed effort, whereas epigenetics is, strictly speaking, about the completion of one's genetic makeup, willed or otherwise, which at birth is still not fully formed. In this respect, epigenetics challenges an assumption shared by Lamarck and Darwin – namely, that we are born with a determinate genetic makeup. This shared assumption underwrote Galton's 1874 christening of 'nature' and 'nurture' as the two independent variables involved in an organism's development (Renwick 2012: chap. 2). However, epigenetics would have us revisit the 18th–19th century debate between epigenesis and preformation as alternative accounts of such development (Moss 2003). Back then the concern was less with how an organism might overcome its genetic load than how it constitutes a relatively open or closed system. From this standpoint, *transhumanists* differ from *posthumanists* in their willingness to engage in epigenetic interventions to reach a desirable closure to the genome's makeup, whereas the posthumanists doubt the long-term efficacy of such efforts at strategic closure. For them all living systems are irrevocably open. In that respect, epigenetics appears exciting because it might advance the transhumanist agenda by enabling the preformationist perspective to be simulated within an epigenesis framework, as the cosmic designer comes to be internalised as a feature of the environment to which an organism is exposed during its development, namely, the 'soft eugenicist' who applies gene therapy via, say, DNA methylation.

This is a good point to turn to the specifically *extensive* challenges to our well-being. These involve the prospect that a wider range of beings

DOI: 10.1057/9781137277077

may be incorporated into society's welfare function in the future – not only non-human animals but also non-animal humanoids. At the outset, it is worth recalling that classical definitions of a liberal society presuppose a set of beings, clearly marked as humans, whose capacity to affect each other is roughly the same. Thus, the ideal of such a society is often said to involve everyone enjoying the most jointly realisable freedom. In other words, I am allowed to do whatever I want as long as it does not interfere with your ability to do likewise. The ideal has been traditionally thought workable because, in the end, however much we may differ in our ends, the means at our disposal are limited to what we make of the bodies of our birth and those of consenting others. Moreover, the fact those means are roughly the same, finite and focused on certain basic common wants and needs provides the ontological framework for both sociality and tolerance.

To see the nature of the extensive challenges to this liberal ideal, consider two senses in which societies might be judged in terms of whether they enable individuals to flourish as human beings. The usual way focuses on human *potential*, namely, the opportunities people are given to exercise their talents or 'capabilities', to recall the version of popularised by Amartya Sen and Martha Nussbaum (Nussbaum and Sen 1993). To be sure, there is no predetermined sense of what people will do with those capabilities, but sheer possession of them constitutes a form of natural capital that deserves to be exploited to increase the world's overall good. A rather different way of evaluating the level of humanity in a society is in terms of its overall ability to bring about human-like or humanly relevant effects, regardless of how or by whom they are achieved. In that case, sheer possession of a human body may not be necessary or perhaps even sufficient for enhancing a society's humanity. Rather, one might think of computer avatars, android companions or even some pets as *virtual* humans who, in virtue of the networks they form with other actual humans, might be counted as functionally more 'human' than other actual humans.

The politics of the potential–virtual distinction are in practice and quite subtle, but the differences in their theoretical starting points are clear enough. In terms of our earlier discussion, the potential/virtual distinction tracks the epigenesis/preformation debate, if we imagine the latter binary as referring to entire populations rather than individual organisms. The distinction is ultimately grounded in alternative ways of glossing the expression 'being human': The politics of potentiality leads

DOI: 10.1057/9781137277077

to an emphasis on what it means to *possess* humanity (and hence lends itself to talk or 'rights' and 'opportunities', both of which preexist any action taken by the candidate human), while the politics of virtuality pushes towards what it means to *produce* humanity (and hence lends itself to talk of 'recognition' and 'outcomes', both of which are consequent on what others make of the candidate humans). In short, the distinction turns on the difference between 'having' and 'doing' one's humanity. Separating them is a difference in temporal horizons – a present that points to, respectively, the past and the future. Each has been compelling in formulating the ideal of a 'free society': On the one hand, we might want to live in a society that enables those marked as 'humans' from the outset to do whatever they want, regardless of consequences. On the other, we might want to live in a society composed of those that, no matter their material (including biochemical) origins, enhances the sense of humanity with which they collectively identify. Let us take each in turn.

The former society – of the 'potentially' human – would be constituted by individuals who already at birth are sufficiently similar to allow for mutual toleration of whatever they happen to do. This line of thought, crucial to the 'rights revolution' of the 18th century, may prove to be the most persistent residue of the biblical idea that we are all descendants of Adam. (John Locke's version of Christian dissent would be the place to begin pursuing this strand.) In Isaiah Berlin's (1958) influential terms, it provides a basis for 'negative liberty', whereby the just society is defined in terms of those willing and able to absorb the consequences of each other's actions. In contrast, the latter society – consisting of the 'virtually' human – understands 'humanity' mainly as an end-state or a 'work in progress', which implies both a greater tolerance for the diverse origins of its prospective members and a more exacting sense of their permissible outcomes. Such a society's would-be members would be burdened with showing that, whatever their material makeup, they are nevertheless making a recognisable contribution to the collective human project. Underwriting this conception is what Berlin called 'positive liberty', which during the Enlightenment was marked by a propensity for crypto-preformationist accounts of human progress by stages (e.g. Condorcet, Turgot) that then persisted for another two centuries in various visions of socialist progress (e.g. Comte, Marx).

To be sure, a liberal society may combine elements of the potentially and the virtually human. To recall a point especially driven home by Peter Singer (1993), even if societal membership is defined in terms of

DOI: 10.1057/9781137277077

possession of the relevant capacities, those capacities are salient not simply because *Homo sapiens* happens to be born with them but because they provide the means for leading a meaningful 'human' life. This suggests a more 'virtually human' welfare orientation, which in turn may require that humans make room for sufficiently 'capable' animals that could also lead such lives. Arguably this strategy was at play in the 19th century as women turned to the justice implicit in animal societies to demonstrate, by contrast, their 'inhumane' treatment in human societies (Bourke 2011). In such cases, the potentially human would seem to verge into the virtually human, such that once women secured legal rights, the idea of 'animal rights' acquired a literalness previously lacking. Moreover, much of the excitement surrounding the recent emergence of epigenetics, noted above, is the prospect of a 'soft eugenics' programme that takes advantage of our genetic plasticity even after birth (e.g. treatments that switch genes on and off) to enable a group of beings – perhaps human and animal – to live in a more mutually compatible world.

But even before the therapeutic virtues of epigenetics have been proven, one practical political context in which the virtualisation of the human may well have some purchase in the future lies in the rationing of healthcare in social security systems associated with the welfare state. At that point, 'Humanity 2.0' is no longer a science fiction slogan but an explicit policy agenda. For example, we are mentally prepared to extend the idea of 'health' beyond the capacities of the normal human body. Indeed, we are prepared to extend it in two rather different senses. To see what I mean, imagine a two-question survey that might be conducted on the future priorities for provision by, say, the UK's National Health Service (NHS):

1   By responding to this survey, you will be covered by the NHS. Now name your two most 'significant others' who reside in this country and whose basic health needs should be covered by the NHS. You are not limited to humans in your answer (i.e. animals and androids may be named).

2   Various mental and physical 'enhancements' are regularly introduced into the market, subtly altering our default settings for normal health. At what level of market saturation for these products should the NHS make them as readily available as eyeglasses and hearing aids?

Of course, we live in tight budgetary times, but even in the best of times we would be unable to honour everyone's requests. So on what basis do

DOI: 10.1057/9781137277077

we make choices? To be sure, there are many imponderables but I would like to propose the following hypotheses.

In the case of the first question, *Homo sapiens* would constitute less than 100 per cent of the list of significant others; perhaps significantly less, if people take the question seriously. This suggests that we need to refocus not only health provision but also medical research – as well as open up medical research budgets to, say, engineers in the business of repairing and enhancing androids – as well as retooling humans as androids. The prospect of blurring the boundary between biology and technology in just this sense had been anticipated in the 1960s by the US Air Force medical doctor, Jack E. Steele, who invented the field of 'bionics' and inspired the hit 1970s television series, *The Six Million Dollar Man*.

In the case of the second question, the various enhancements would alter our sense of what it means to live a fulfilling and meaningful human life. There may be some unexpected and even perverse results. Would the overall effect be to assign less value to the lives of those who, by choice of by fate, are unenhanced or unenhanceable? Moreover, some may wish to enhance their animal and android companions, whilst others may wish to turn 'enhanceability' into a threshold for a fulfilling life, below which public health provision may be withdrawn. The two wishes might even work in concert to raise the moral status of some animals and androids above some humans. Peter Singer opened the door to this way of thinking when he proffered sentience as the threshold for moral relevance in defence of animal welfare.

The NHS turns out a useful concrete site for thinking about Humanity 2.0 because, as the cornerstone of the UK's welfare state, it gave a very clear sense of the quality of life to which everyone was committed on behalf of everyone. The 'everyone' of course was understood to be all and only members of *Homo sapiens*, the vast majority of whom would contribute to the funding of the NHS through their taxes. Humanity 2.0 is about possibly redrawing that boundary and all the implications this has for health policy, public policy more generally and broader social and economic relations. If you still think that it is premature to take these matters seriously, keep three considerations in mind.

1 Although Humanity 2.0 is still in its infancy, people are already voting with their feet to get into it. The best indicator is the increasing amount of time that people spend in non-face-to-face, non-human communication. This trend may be most easily seen in the increasing

DOI: 10.1057/9781137277077

number of single-person households. But even in more 'normal' social arrangements, the time spent both with animals and in front of computer-based devices implies a radical, albeit relatively quiet, transformation of the terms in which the bonds of our social life are being forged. One does not need to join in the jeremiads of Sherry Turkle (1984) and Susan Greenfield (2003) to think it is probably true that our cognitive and emotional ties are in the process of substantial rewiring.

2 Both public and private agencies are devoting increasing resources to the 'anticipatory governance' of Humanity 2.0. This involves inviting people to test-drive innovative lifestyle-changing goods and services by participating in focus groups, citizen juries, scenario construction, wiki media and virtual reality (Barben et al. 2008). The underlying principle here is that any misgivings that people might have about such innovations – whatever their basis – may be rectified before the products come on stream. In any case, people will have begun to expect the regular appearance of, say, 'enhancement' technologies, and may even call for them to come sooner.

3 The unravelling of the social contract that underwrote the NHS and the rest of the post–World War II welfare state, which presupposed the clarity and integrity of Humanity 1.0, is not an isolated event. It reflects the failure of modern political ideologies to capture the imaginations of the vast majority of people, not least the young. Yet, the resurgence of fundamentalist movements focused on race and religion points to an appetite for rethinking the boundaries of social and moral concern. At the same time, the ease with which people can opt out of any collective engagement with social welfare issues – the various flavours of 'privatisation' on offer today – suggests that a strong political vision will be needed to ensure that the identity of Humanity 2.0 does not simply turn into a perverse aggregate effect of many narrowly self-interested decisions.

DOI: 10.1057/9781137277077

3
Anthropology for Humanity 2.0

Abstract: *The discipline of anthropology has been traditionally and unapologetically dedicated to 'Humanity 1.0', which corresponds to the default normative expectations of beings born with our bodies. However, we have always been pulled in two different directions: a transhumanist direction that would have us transcend nature, and a posthumanist one that would embed us more in nature. The former, which I broadly support, involves a desire to intensify and extend uniquely human properties beyond their normal physical parameters, whereas the latter, which I broadly oppose, involves an indifference, if not hostility, to the original humanist project. My paradigm case for each in today's world is, respectively, Ray Kurzweil and Peter Singer. I explore their respective implications for concrete living conditions. Together the two cases make clear that 'Humanity 2.0' is about the matching of function to form or message to medium. Finally I show how these prospects reflect an understanding of science as a kind of 'artificial theology' (in contrast to 'natural theology'), the process by which through the continual transformation of world and self, we come to realise the divine potential of our being.*

Fuller, Steve. *Preparing for Life in Humanity 2.0.*
Basingstoke: Palgrave Macmillan, 2013.
DOI: 10.1057/9781137277077.

The discipline of anthropology has been traditionally and unapologetically dedicated to 'Humanity 1.0', which corresponds to the default normative expectations of beings born with our bodies. However, the existential trajectory of this being has always been pulled in two opposing directions: on the one hand, a 'posthumanist direction that would embed us more in nature; and on the other, a 'transhumanist' one that would have us transcend nature. Indeed, I draw a sharp distinction between *posthumanism* and *transhumanism*: The former, which I broadly oppose, involves an indifference if not hostility to the original humanist project, whereas the latter, which I broadly support, involves a desire to intensify and extend uniquely human properties beyond their normal physical parameters. Starkly put, posthumanism is *anti-humanist*, while transhumanism is *ultra-humanist*. My paradigm case for each in today's world is, respectively, Peter Singer and Ray Kurzweil. In this chapter, I explore their respective implications for concrete living conditions and educational requirements. Together the two cases make clear that 'Humanity 2.0' is about the matching of 'function to form' (biology) and 'message to medium' (technology). Finally I show how these prospects reflect an understanding of science as the culmination of *theosis*, the process by which through a continual transformation of world and self, we come to realise the divine potential of our being. In this context we might speak of an 'artificial theology' of the future that revisits in a new key the central problems of 'natural theology', especially the respective roles of faith and reason in the worldview of Humanity 2.0.

3.1   Two ways of getting over the human condition

I recently gave a talk entitled 'Who will recognise "Humanity 2.0" – and will it recognise us?'. Although the title struck some as facetious, it was meant as a reminder that any account of 'getting over' the human condition implies certain views about where we came from, where we are going and, perhaps most importantly, how we and some successor species might relate to each other. For example, Darwinian evolutionary theory may be understood as positioning us as 'Primate 2.0', in which humans and simians are expected to understand each other's lives only in rather limited but (hopefully) benign ways. To think of 'Humanity 2.0' along these lines is characteristic of *posthumanism*. The posthumanist 'gets over' being human by seeing us as just one among many species past, present

DOI: 10.1057/9781137277077

and future – all equally valued but none valued unconditionally. Each species is entitled to live the fullest life it can to the fullest extent it can. At first blush, posthumanism speaks to an eco-friendly liberal sensibility. On closer inspection, it suggests that humanity's celebrated cerebral cortex may be little more than an anthropocentric conceit that will prove to be our undoing at the hands of natural selection. Those who would place global climate change at the top of the political agenda intuitively resonate to posthumanist concerns. Posthumanists would blame human hubris for this dire state-of-affairs, which they would in turn trace to an exaggerated faith in the powers of science, which *Homo sapiens* have developed into a secular religion. Yet, exactly this provides the springboard for an alternative sense of 'getting over' humanity, namely, *transhumanism*.

Transhumanists unabashedly privilege humanity above other animals– but without making us the end of the story. Science will find ways to leverage the distinctly human out of the messy wetware from which we have evolved. Just because it has taken billions of years working through carbon-based organisms to get to where we are now, it does not follow that we need to continue along the same lines. Indeed, a defining feature of our humanity is that we came up with theories like Darwin's, which enable us to understand evolution as a whole, so that we might steer its course in the future. When Julian Huxley (1953) first developed transhumanism as an idea in the middle-third of the 20th century, he was mainly thinking in terms of the kind of genetic stewardship associated with eugenics. Nowadays transhumanism concentrates less on features of humanity that should be 'selected out' than on which should promoted or enhanced. While some transhumanists follow Aubrey de Grey (2007) in continuing to envisage their project in terms of extending indefinitely the survival of the human body, others are more inclined to Ray Kurzweil's (1999) view that not everything about our biological character is worth preserving – especially if humanity is strongly identified with our mental properties, which would favour the sort of carbon-to-silicon migrations traditionally found in science-fiction stories of cyborgs, androids and avatars. Transhumanist tales – both fictional and futurist – usually involve a mutual understanding that we and our successor species are part of the same line of descent, which for better or worse enables us to see them as more closely realising our own potential. While flattering on one level, it can easily lead our transhuman successors to discard us like old drafts of a now more-finished plan. In contrast, our posthuman successors might leave us alone in benign neglect!

DOI: 10.1057/9781137277077

Four questions may be asked to distinguish *posthumanists* and *transhumanists*:

1 *What is your default attitude towards humanism?* Do you believe that it was always a bad idea, the narcissistic ideology of an elite that only served to alienate and subordinate not only most so-called humans but also the rest of nature? Or, rather, do you believe that humanism's historic excesses merely reflect the arrested development of a fundamentally sound ideal that will be more fully realised once we manage to reorganise ourselves and the environment, perhaps involving a scientifically informed sense of global governance?

2 *What is the source of the conflict between science and religion that has been so prominent in the modern history of the West?* Do you believe that it lies in the Abrahamic religions' continued privileging of the human as the locus of meaning in nature, despite the successive displacements of human uniqueness wrought by Copernicus, Darwin and Freud? Or, rather, do you believe that the conflict is simply a reflection of science's increasing capacity to deliver on the promises of the Abrahamic religions by demonstrating how we might turn nature, including our own animal natures, to realise our divine potential?

3 *How do you see Homo sapiens in the context of evolutionary history?* Do you believe that the highly developed cerebral cortex that marks our species is little more than an entrenched genetic quirk that will eventually undermine our descendants, once the environment turns that quirk into a liability? Or, rather, do you believe that the special powers enabled by our cerebral cortex provide an opportunity for us to break free from biological evolution altogether, say, by developing means for our uniquely human qualities to migrate from their current carbon-based containers to more hospitable, possibly silicon-based ones?

4 *Is there some normatively desirable intentional relationship in which a successor species should stand to us and our ancestors?* Do you believe that evolution is so blind that it would make as little sense for a successor species to think of itself as improving upon what we had been trying to do as for us to think that we are advancing projects initiated by our simian ancestors? Or, rather, do you believe that the legitimacy of any future transformation of the human condition, however fundamental, depends on the counterfactual prospect that we and our illustrious

DOI: 10.1057/9781137277077

ancestors would recognise it as contributing to realising our deepest aspirations?

In response to each question, *posthumanists* take the *former* option, *transhumanists* the *latter*. The differences may be epitomised as one of historiographical perspective. The 'transhumanist' supposes that life is purposeful not simply at any given moment for a given organism but in its entirety – it has an overall direction. In this context, 'humanity' is the name used to personify the leading edge of this overall process. In contrast, the 'posthumanist' supposes that 'humanity' simply names a transient species within life's flow that, to be sure, has its own perspective on the overall process but no special vantage point. As a first pass, this distinction reproduces the two main modern evolutionary frameworks – the teleological Lamarckian one (transhumanist) and the non-teleological Darwinian one (posthumanist).

The above distinction in evolutionary perspectives also resonates with two different thresholds one might set for 'consciousness': *intentionality* (i.e. thought reaching beyond itself to something desired) and *sentience* (i.e. thought as a self-satisfying experience). The former veers transhumanist, the latter posthumanist. However, as the Italian postmodern theorist Giorgio Agamben (1998) has argued, distinctions of this sort have been historically difficult to uphold. In particular, the more one has believed that the purposes of humanity extend beyond living human bodies, the greater the tendency to treat those (including our own) bodies instrumentally and perhaps even inhumanely. Indeed, today's ethical trans-/post-humanist struggles were already prefigured in 16th century diplomatic controversies over actions taken against individuals for *raison d'état* (i.e. the national interest). Both sides of the legal debate claimed to speak on behalf of 'humanity', yet what the proto-posthumanist defenders of the integrity of individual bodies called *torture*, the proto-transhumanist defenders of state prerogative termed *sacrifice*. In this context, Hobbes' *Leviathan* appears as a blueprint for 'Transhumanity 1.0', namely, the corporate android, the literal 'body politic' whose cells, tissues and organs are perpetually regenerated by each new generation of individual humans who agree to be bound by the social contract. Kant may be seen as a Hobbesian who justifies the psychic transformation of successive generations of citizens who internalise the state's perspective as their own, so as to interpret sacrifice as a personal obligation to embrace pain in the name of a higher, if not ultimate, sense of self-realisation.

DOI: 10.1057/9781137277077

In this moment of sacrifice, they experience a virtual freedom that (so they hope) anticipates the actual freedom of others in the future. This certainly explains a lot about the modern German worldview (i.e. from Kant to the Kaiser), in which the sense of belonging to a larger collective enables the individual to experience the world in ways that releases powers in them that would otherwise remain dormant (Krieger 1957).

Today's trans-/post-humanist struggles are perhaps more recognisably traced to the vivisection movement championed by Thomas Henry Huxley in the 1870s on behalf of Britain's emerging biomedical establishment, but opposed by the Society for the Prevention of Cruelty to Animals (SPCA), whose original patron was Peter Singer's spiritual father, Jeremy Bentham. Posthumanists follow the SPCA in taking the metaphysical backdrop of Darwin's original formulation of the theory of evolution by natural selection more seriously than did his great defender Huxley: No matter how much we may succeed in turning other species and the environment to human advantage, we are ultimately overwhelmed by nature. While this perspective, familiar from Epicurus and his followers in the ancient world, enables one to be open to the welfare of other creatures subject to a similar fate and humanity's affiliations with them (including quite unconventional 'hybrid' ones), there is no sense that significant long-term change will come from the sort of concerted human intervention normally associated with implementing a long-term, large-scale plan. Here the denial of human *control* should not be confused with the lack of human *impact*, which posthumanists easily grant, especially when bemoaning 'anthropogenic' climate change.

The emphasis that posthumanists place on Darwin's own understanding of evolution is reminiscent of 'Primitive Christianity', whose legitimacy rests on accessing an 'original' understanding of Christ's message stripped of clerical and theological baggage. The secular analogue of such baggage is an exaggerated faith in our 'biotech century' (Rifkin 1998) born of the history of the biomedical sciences *after* Darwin. *On the Origin of Species*, it should be recalled, was published a half-century before the rediscovery of Mendelian genetics and a century before the discovery of DNA's double-helix structure. Darwin himself was pessimistic about our ability to fathom the inner workings of life and arguably did not believe that life was tractable to mechanistic explanations at all, let alone the sorts of computer simulations of evolutionary history on the cutting edge of Neo-Darwinian research (e.g. Lenski et al. 2003). His own view of humanity's fate was probably rather close to that of the

DOI: 10.1057/9781137277077

late Stephen Jay Gould, whose own research expertise followed Darwin's in stressing the field over the laboratory. To underscore our ephemeral cosmic status, Gould (1988) notoriously claimed that were the tape of evolutionary history replayed, *Homo sapiens* would probably not feature in it at all.

Posthumanists reject the historically most powerful reason for believing that humans might exceed other creatures in their control over nature – namely, the biblical doctrine of *imago dei,* that is, that humans are created in image and likeness of the creative deity. The biblical prerogative is asserted even by Thomas Hobbes in the opening paragraph of *Leviathan,* and this unquestioned adherence to species chauvinism, or 'speciesism', has been invoked by posthumanists to explain the failure of Marxism as a political project (Singer 1999). For the posthumanist, the sense in which 'natural selection' is an extension of 'artificial selection', the means by which humans have domesticated nature, is merely *analogical,* not *univocal* – to recall the distinction that the medieval scholastics drew with regard to the semantics of God-talk (Fuller 2011: chap. 2). In other words, natural selection (formerly God) should not be seen as simply a much more powerful version of our plant and animal breeding practices (i.e. univocally), which might suggest that humans are uniquely placed to acquire nature's own powers in the future. Rather, the way we breed plants and animals gives us an intuitive sense of how patterns of biological descent *might* have arisen in nature – but it does not licence any deeper sense of identity in the causal dynamics of the two modes of selection. Thus, the posthumanist follows Darwin's own lead in questioning whether the surface similarities between artificial and natural selection plausibly imply an overlap between human and divine intelligence (Fuller 2008b: chap. 2).

The increasing salience of the post-/trans-humanist distinction suggests that had Galton not been Darwin's cousin, Darwin might never have been associated with eugenics at all. After all, Darwin coupled an empirically comprehensive account of natural history with a worldview that denies our capacity to direct its future significantly. The lumping together of Darwin, Spencer, T.H. Huxley and Galton as 'Social Darwinists' reflected Ernst Haeckel's omnibus importation of their ideas to Germany, which spread through the 'racial hygiene' movement that flourished in medical schools in the half-century prior to Hitler (Proctor 1988). However, from today's standpoint, we would say that the more *laissez faire* Darwin and Spencer were 'posthuman evolutionists', whereas the more interventionist Huxley

DOI: 10.1057/9781137277077

and Galton were 'transhuman evolutionists' (Fuller 2011: chap. 1). At stake in their late Victorian heyday was the soul of 'liberalism', the political ideology most closely aligned to humanism: Should it stand for nature taking its course or for us to steer an otherwise directionless nature? In the 20th century, the former veered right (e.g. Hayek) to neo-liberalism, the latter veered left (e.g. Keynes) to welfare-statism (Fuller 2011: chap. 5).

Of course, Galton himself did not succeed in mastering the laws of heredity, but 20th century genetics was populated by some rather heterodox Christians who interpreted the charge of 'playing God' as a literal extension of our divinely mandated stewardship of the earth: for example Ronald Fisher, Sewall Wright, and most notably Theodosius Dobzhansky (1967). We might even reach back to Gregor Mendel, the long-neglected Moravian monk who first demonstrated heredity's most basic statistical laws. His mentor was a devotee of the Unitarian preacher and radical chemist, Joseph Priestley, who clearly tied human emancipation from its fallen animal state to our impending control over the forces of nature (Wood and Orel 2005; Fuller 2011: chap. 4). Moreover, the two figures most closely associated with the original mid-20th century usage of 'transhumanism', Julian Huxley (grandson of Thomas Henry) and Pierre Teilhard de Chardin (1961), saw the transition from Christian theology through evolutionary science to an evolutionary humanism as a seamless process of divine self-realisation (Fuller 2007c: chap. 5).

In this context, Jesus symbolises the continuing need to attend to persistent features of the human condition – notably poverty and injustice – that inhibit our divine potential. The difference between Huxley, who called himself a 'humanist', and Teilhard, who called himself a 'Christian', was that Huxley treated universal self-consciousness as an emergent feature of the evolution of matter, whereas Teilhard retained the Abrahamic idea that God preexists matter but creates by steering the evolutionary process. But in both cases, each human being is presumed to be of value as a unique participant in a superorganism, typically modelled on a world-brain, what Teilhard called the 'noösphere'. The idea was already present in Condorcet's defence of increased human population growth as a vehicle for the 'hominisation' of the world, the ultimate materialisation of Enlightenment ideals (Fuller 2006a: chap. 13). Historians have tended to treat this continuity between Christianity and the Enlightenment as a mere atavism, what Carl Becker (1932) famously called 'the heavenly city of the eighteenth-century philosophers'. However, it is perhaps better understood as a moment in Christianity's evolution to transhumanism,

DOI: 10.1057/9781137277077

in which Hobbes' absolute sovereign, Comte's positivist church and perhaps even the Marxist ideal of a global Communist order also serve as way stations.

Clearly, unlike the posthumanist and much more than the classical humanist, the transhumanist sees our overdeveloped cerebral cortex – the seat of intelligence – as less an biological accident than a harbinger of things to come. In the transhumanist case, it might even involve abandoning biology for technology as the relevant medium for our evolutionary future. But how are we to understand this promised advance in intellectual powers? Without resorting to science fiction, the most natural place to look is to the self-transcending forms of cognition that have been put forward by philosophers and theologians, especially when defining the overlap between human and divine being. Perhaps the clearest secular-philosophical residue of this line of thought is the so-called *a priori* knowledge, which supposedly enables us to grasp basic structural features of reality with relatively minimal sensory input, as if our being were not confined to our animal bodies.

Great claims have been made about the reach of such knowledge, usually inspired by the universal applicability of mathematics (Wigner 1960). Thus, Leibniz held that through 'intellectual intuition' one might infer the laws of nature by feats of heroic abstraction tantamount to adopting God's standpoint, the proverbial 'view from nowhere'. Kant famously debunked such alleged powers of mind as empirically unsubstantiated assumptions that nonetheless may be psychologically required to underwrite our continuing faith in science. However, Kant's strictures were rendered obsolete within a century as the empirically unrestricted mathematical imagination managed to anticipate forms of experience (e.g. non-Euclidean geometry, transfinite set theory) that resulted in revolutions in physics and logic in the early 20th century. Moreover, the increasing migration of scientific inquiry to the virtual reality of computer simulations in the late 20th century adds to the sense that perhaps our capacity for intellectual intuition needs to be taken more seriously (cf. Horgan 1996).

As we saw at the end of the last chapter, health policy provides a prism from which to witness the different evolutionary assumptions of post- and trans-humanists in action. Humanists, influenced by the Abrahamic privileging of our species being, have supported scientific research and clinical practice designed to keep the entire human body functioning as long as possible, regardless of cost. This principle has been applied

DOI: 10.1057/9781137277077

across the age spectrum, from the unborn to the very old, including the multiply disabled. In this respect, humanism is ideologically promiscuous: Both the Left and the Right can claim its arguments. The same promiscuity applies to posthumanists, who are more concerned with the promotion of life as such than human life specifically. They have supported the premodern medical aim of a 'good death' (once a 'good life' can no longer be led) as well as the modern aim of life's efficient servicing by treating everyone as potential 'organ donors', perhaps even to other species, where appropriate. The practical implications of the posthumanist ideal – perhaps with a jibe at Peter Singer – are vividly on display in 'Utilitaria', one of the destinations of the fictional Professor Caritat (Condorcet's family name) in Steven Lukes' (1996) recent update of *Gulliver's Travels*. For their part, transhumanist visitors to Utilitaria, a hybrid of Britain and Sweden, would be less troubled by the radical utilitarian framework for welfare provision, with its apparent disregard for the integrity of organisms, than Utilitaria's normative commitment to a steady-state ecology whereby life is simply recycled but never improved.

In contrast, a good entry point into transhumanist health policy is to recall the special value attached to human intelligence. Thus, a premium would be placed on preserving a recently deceased human brain, perhaps cryonically or attached to some suitable electro-chemical medium, for future resurrection. In the long run, transhumanists would like to overcome the leakage problems of 'wetware' altogether by enabling one's worldview, if not personality, to be uploaded to a silicon chip that is then implanted in an android. Once again, they prove themselves to be much less carbon-fetishists than either humanists or posthumanists – and hence happier to deviate from the Darwinian sense of biological evolution. (I explore the implications of this point in the Epilogue.) Indeed, transhumanists would prefer to keep the DNA of various species, including *Homo sapiens*, on tap for discretionary use than sustain the fully matured bodies of such species in well-policed ecologies. In the end, these ecologies amount to no more than global projections of zoological and botanical gardens – exactly how Linnaeus originally saw them in the 18th century, when promoting their construction in order to reinstate humanity's biblical prerogative, which he took to involve recovering the conditions that had obtained in Garden of Eden (Koerner 1999). While today's transhumanists can hardly object to Linnaeus' anthropocentric biosphere, as beneficiaries of subsequent developments in genetics and molecular biology, they are more likely to look to the storage facilities of laboratories than outdoor gardens, however

DOI: 10.1057/9781137277077

well-managed, to house organic paradigms for technologies that enhance our own capacity for survival and extension – a field nowadays known as 'biomimetics' (Benyus 1997).

Finally, a key site for exploring the transhumanist evolutionary sensibilities is so-called interspecies research (ISR), which ranges from the implantation of human stem cells in a mouse embryo to the transplantation of organs from a genetically modified pig into a human body. In the former case, the mouse is used as a breeding ground for human organs; in the latter, the pig is actually bred to provide those organs (Bonnicksen 2009). At first glance, both humanists and transhumanists can support ISR because of the priority it gives to the promotion of human welfare. But humanists would be less eager than transhumanists to license an open-ended exploratory approach to transgenic organisms, which may result in new living beings, so-called 'chimeras' and 'hybrids' that threaten to impose new ethical burdens on humans. Such qualms would not arise for transhumanists, who take a more instrumental attitude towards animal embodiment, reflecting a generally demystified attitude towards the very idea of an organism. However, the prospect that these transgenic entities might be ends in their own right – and not mere means to human ends – engages posthumanists in the moral discussion surrounding ISR, in which they become the protectors of 'life itself'. One prospect on the horizon that would bring together transhumanist and posthumanist conceptions of evolution is the rise of 'systems' or 'living' architecture, in which organisms are embedded – and in some cases engineered – to complement the life cycle of ordinary construction materials, so as to provide for more ecologically integrated buildings capable of literally repairing themselves with minimal human intervention (Armstrong 2012). This would be an instance of a posthumanist sensibility embedded within a tranhsumanist one that at least in the short-term supports our lingering commitment to humanism. Let us now turn to consider this aspect – designing living spaces for Humanity 2.0.

3.2　Designing living spaces for Humanity 2.0

Even though it required the United Nations Declaration of Human Rights in 1948 to make explicit 'Humanity 1.0' as a set of universally binding normative standards for the treatment of *Homo sapiens*, a persistent feature of the laws enacted on behalf of Humanity 1.0 over

DOI: 10.1057/9781137277077

the previous three centuries – that is, since the Peace of Westphalia – is both their protection of individual humans against each other and their empowerment of all humans over nature. (The former is the liberal moment of modern civil law, the latter is the socialist moment.) Humanity 1.0 regimes may operate democratically or not, but they all license the massive reengineering of the physical environment to enable beings of that sort to flourish. The results have been, in equal measure, impressive and precarious. For 50 years now, one or another ecological crisis has been declared, most recently in the name of global warming. In each case, the indefinite extension of Humanity 1.0 has had to shoulder the blame.

What distinguishes the current wave of ecological crises is neither their severity nor their urgency – nor even the certainty that they will come to pass. Rather, it is the breadth of options available to deal with them. To be sure, many of the options are more conceptual than material, but all involve renegotiating the role of 'form' and 'function' in the human condition. In particular, the hard ontological boundaries between 'human' and, on the one hand, 'animal' and, on the other, 'machine' are dissolving. This is not necessarily because all of these entities are being treated as equals; rather, they are increasingly treated as overlapping in form and interchangeable in function. In short, we need to envisage a world in which humans, animals and machines inhabit a common social ecology, learning from each other as they forge a mutually sustaining existence.

However, to reach this new understanding, we first need to see how 'Humanity 1.0' kept them apart. Karl Marx was a classic Humanity 1.0 thinker. Humanity 1.0 puts peoples in houses, machines in factories, and animals most everywhere else, though they have been increasingly constrained and organised by the imperatives of house- and factory-building. Here Marx divined the logic of capitalism: After all, what is not dedicated to the production (factory) or reproduction (house) of human labour is nominally left to the animals – that is, until they turn into pets or food. The three sorts of entity are presumed to be mutually exclusive, existing at best in complementary – and often competitive – relations with each other: One provides what the other cannot, but they are not the same and one cannot turn into the other.

In this context, a good way to think about Humanity 2.0 is as questioning the negotiated settlement between *form* and *function* that has left an indelible stamp on the planet, originally to mark the progress of our

DOI: 10.1057/9781137277077

species but increasingly to threaten our survival. I deliberately refer to 'form' and 'function' when describing the human condition, since the binary is invoked just as easily when speaking about the makeup of the artificial and the natural world. In the world of artifice, a stress on 'form' has been associated with an 'aesthetic' worldview, a stress on 'function' with an 'instrumental' one. In the world of nature, a stress on a 'function' has been linked to a teleological (Lamarckian) view of evolution, a stress on 'form' with a non-teleological (Darwinian) one (Fuller 2007c: chap. 1). Humanity 2.0 entails the convergence of the natural and artificial senses of 'form' and 'function' as we come to think of ourselves as *deities on a budget* – that is, as engaged in ever more scaled-up micro-versions of the task that faced the Abrahamic God when imposing a design on recalcitrant matter in the course of creating the world. This direct comparison between God and humans in terms of the trade-offs that are required for each to realise their respective creative capacities is the stuff of *theodicy* – the branch of theology concerned with the nature of evil in a supposedly good world (Fuller 2010a: chap. 7).

Deities on a budget think of organisms as the outcomes of an economic production process, in which formalists impose a supply-side regime and functionalists a demand-side one. A good way to think about this budget is as posing in a new key, the question asked by modernist movements in 20th century art: *Does function follow form or form follow function?* In short, an important consequence of the increasing interaction amongst humans, animals and machines is that their identities have been blurred. More specifically, as humans have come to identify with animals, *function has followed form*, whilst as they have identified with machines, *form has followed function*. Let us take each proposition in turn.

Function follows form: There is a 95+ per cent genetic overlap between humans and other animals. Evolutionary precedents have been found for many cognitive and emotional traits that were previously considered uniquely 'human'. These facts have emboldened 'animal rights' activists to declare the prospect of animals as either pets or food equally horrific. Moreover, the same facts are causing humans to recalibrate their sphere of care and concern in 'posthumanist' ways. Perhaps health insurance should be extended to animals that live in our midst? Taking seriously this newfound dignity to animals would involve redesigning spaces – not only homes but also clinics and courts – to respect the physical differences between humans and animals that would remain, but would now no longer make a normative difference. One obvious change would be an

DOI: 10.1057/9781137277077

increased tolerance for mess and waste, which anthropologists have long pointed to as an important means by which humans keep their distance from animals – not least other humans who 'behave like animals'.

Form follows function: From their inception, machines have inspired a love-hate relationship in humans. In cultures touched by the Abrahamic religious tradition, where this feeling has probably been strongest, it is traceable to the reminder that humans themselves are created 'in the image and likeness of God'. After all, machines are typically designed to excel at activities that humans value, or at least find relevant to the maintenance of their existence. In principle, of course, these artificial creatures should free-up our lives to do other things. But in practice, as Marx emphasised, the machines might overtake and replace humans – if not exterminate them altogether, as in so many science fiction dystopias. But Marx did not anticipate that some machines – computers – might be created that closely capture and amplify many, even if not all, of our mental powers, which then serve as platforms for launching alternative 'virtual' mode of beings, with which users over time come to identify more strongly than with their own bodies. A sign that we have already taken significant steps in this direction is the increasing compatibility between sustained and sophisticated interactions with digital media and their users' disregard for the state of their own bodies and the immediate environment.

These two vectors of Humanity 2.0 are clearly countervailing, yet both are on the rise, ever more pronounced amongst younger members of the population. Thus, spaces need to be designed that accommodate both projected futures: One that enables us to reconnect to our animal nature, and the other where that biological baseline is treated as merely a vehicle to journey into a more fulfilling cyber-existence. Once these parameters are in place, we can easily imagine intermediate positions: for example, prosthetically enhanced humans, chimeras and hybrids that combine human and non-human materials, and even molecular and digital infor-mation that is copied and preserved for purposes of future uploading in new 'media', in a sense that will be increasingly indifferent to whether one is talking about an organism or a machine.

An example related to the last, perhaps most outlandish possibility is already being developed by Rachel Armstrong (2012) and her colleagues at Greenwich University in London as 'metabolic materials'. These are based on 'protocells', chemical substances that can be programmed to respond sympathetically to buildings, offering them support in the face

DOI: 10.1057/9781137277077

of changes in the physical environment. Armstrong, whose own hybrid existence combines medicine, science fiction and architecture, argues *inter alia* that such a reengineering of nature gives Venice its best chance of not sinking into the sea. At the moment, her work is just as likely to be exhibited as an instance of synthetic biology, a piece of innovative technology and an art installation. In the world of Humanity 2.0, the glass separating those display cases is bound to be broken as well.

3.3 Science as the theology of Humanity 2.0

Science and technology would not be as they are today, were it not for 'Abrahamic theology' – that is, the common core of Judaism, Christianity and Islam, which the German Enlightenment critic and dramatist Gotthold Lessing canonised as 'monotheism'. Lessing had at least two things in mind, when he lumped together religions which historically had expressed considerable mutual hostility: (1) the idea that humanity is uniquely created in the image and likeness of God, which means that, for better or worse, we are potentially capable of both godlike knowledge of what can happen and responsibility for what does happen; (2) the idea that language is the privileged medium through which God communicates to humans, which relates to (1) insofar as the more language-like that reality appears, the more our divine capacities are vindicated. This was also the subtle message of Erwin Schrödinger's 1943 Dublin lectures, published as *What Is Life?*, which inspired the younger generation of scientists in his audience to 'crack' what we now routinely call the 'genetic code' (Schrödinger 1955).

The person responsible for seeing science and theology so mutually implicated in their fates in this modern sense was none other than the founder of the scientific method, Francis Bacon, who served as lawyer to King James I when he commissioned the first English Bible. In *The Advancement of Learning* (1605), Bacon proposed that the divine plan is inscribed in two books – the Bible and Nature itself, the former declaring what God intended (i.e. his 'will') and the latter what God has actually produced (i.e. his 'power'). Bacon emphasised that God is the source of both but that they are different. Here Bacon is most naturally read as saying that God wants us acknowledge that reality is not yet as he has designed it to be, which means that there is scope for humanity to redeem itself from Original Sin by fully realising God's intentions under

DOI: 10.1057/9781137277077

these compromised circumstances. In this respect, the ontological distinction between 'is' and 'ought' is simply how we represent to ourselves the distance to divine salvation (Fuller 2011: chap. 2).

It was also during this period – corresponding to the peak of the Protestant Reformation – that 'literal' readings of the Bible were popularised. But there are two senses of 'literal' here, which roughly correspond to the distinction that the late 19th century Neo-Kantian philosopher Wilhelm Windelband drew between *idiographic* and *nomothetic* modes of inquiry. The more easily recognised sense of 'literal' is idiographic. It involves reading the Bible as a straight historical narrative, which (based on the stated age of death of the patriarchs) implies that God created the world in 4004 BC. Modern science has clearly disavowed this reading. However, the Bible may also be read 'literally' in the nomothetic manner of a scientific theory – that is, as an abstract model of recurrent patterns in the world, which remain relevant until the end of time. This is the sense in which we read 'literally' Newton's equation 'F = ma' (i.e. the product of body's mass and acceleration constitutes the physical force exerted by the body).

Sometimes theologians call this latter way of reading the Bible 'allegorical', but it downplays the reading's specifically *performative* character (Fuller 2008b: chap. 7). After all, 'F = ma' is not a relationship that can be observed with the naked eye, as one sees shapes in clouds. (Otherwise, Aristotle would have seen it!) Rather, it involves dealing with the world as if the equation were true but given the resources at one's own disposal. In that case, intellectual intuition 'anticipates' sensory observation in the quest of fully realising the target truth. The next step is to construct a public demonstration to make plain the concrete implications of 'F = ma'. That construction is an experiment, which (if compelling) can serve as a standard for evaluating and reconstituting the world. This is the deep point behind the claim that engineering is an 'application' of physics – and, more generally, technology is an 'application' of science. Even if, as Nancy Cartwright (1983) famously maintains, scientific laws are not strictly true, they nevertheless carry normative force that end up *making* them true, as people realise the benefits of treating the world as if 'F = ma' were true. This is what is literally meant by 'spreading the benefits of science and technology throughout the world', a process intimately tied to our identity as 'modern' and 'progressive' *secular* beings.

Of course, performative readings of the Bible have extended beyond science to politics, most notably in the history of the United States. To be sure, the original 17th century Puritan settlers, the 18th century Founding

DOI: 10.1057/9781137277077

Fathers (who were Deists and Unitarians), as well as the 19th century Christian Scientists and Mormons, were all religious dissenters who were more comfortable with science than with the Christian churches of their day. Nevertheless, they all read the Bible as a document that empowered them to realise the Word of God in their lifetime. However, theologians today tend to underestimate the significance of treating the Bible as these people did – namely, as a dramatic script authored by God yet enacted by us. In this respect, theologians need to recover the sense of *theos* behind the activities associated with *theatre*. The American dissenters took quite literally the proposition that we can all be like Jesus by trying to find inside ourselves the moment of *theosis*, the so-called 'Transfiguration' recounted in the New Testament, when Jesus manages to adopt, at least briefly, God's point-of-view. This orientation should be familiar from the influential 'system' or 'method' approach to acting pioneered in the early 20th century by the great theatre director Constantin Stanislavski, who called himself a 'spiritual realist' (Benedetti 1982).

Curiously, today 'theology' seems to be about everything *except* the literal 'science of God'. Indeed, a popular interfaith position is that of *apophasis*, the fundamental ineffability of our experience of the divine: At most we can know God in terms of what he is *not*. This position is not simply a counsel of humility and tolerance, but it also defeats all highly discursive forms of religious expression, not least natural theology. Typical in this regard is Karen Armstrong's (2009) recent popular book, which argues that religion took a seriously wrong turn when it combined with science as 'natural theology' in the 17th century, from which she traces much of the fanaticism associated with both religious fundamentalism and scientific creationism.

My own conclusion is the exact opposite of hers. Modern science is a natural development of Abrahamic theology, if one operates with a sufficiently broad notion of reading the Bible 'literally', which is to say, not merely as historical description but also as dramatic paradigm. Indeed, the latter reading is arguably superior at a theological level, as Jesus periodically made the point that we should inhabit the Word of God as if we had been its authors rather than simply follow it in the manner of children obeying parents. (This point should be recognisable from Kant's categorical imperative and his insistence that 'Enlightenment' means humanity's freedom from a state of parental dependency.) Seen in this light, the modern scientific impulse to conceptualise physical reality in terms of a set of overarching laws and the modern political impulse to

DOI: 10.1057/9781137277077

organise society according to a set of constitutional principles – both of which are rightly seen as 'foundationalist' – should be seen as concrete attempts to enact the idea that we have been created 'in the image and likeness of God' and hence are capable of imposing an intelligent design on the world. It follows that our successes and failures in these matters provide important clues to the *modus operandi* of God's agency – as well as our distance from fully comprehending it. This should be the main concern of a discipline that dares to call itself 'theology', on the basis of which a science of 'Humanity 2.0' could be built. Such a science would inform the technologies by which we become one with God (Noble 1997).

In conclusion, Humanity 2.0 provides an opportunity to make explicit the sort of entity that a 'scientist' must be in order for science to prove successful. But doing this requires taking literally the seemingly innocent metaphors of 'getting into the mind of God' (physics) and 'playing God' (biomedicine). Notwithstanding scientists' own disclaimers, as a matter of fact science has achieved as much as it has because scientists have adopted a 'godlike' attitude toward nature. This stance constitutes a special kind of subjective agency, one that is typically masked in characterisations of the scientific method that attempt to make the original researcher irrelevant to the evaluation of her knowledge claims. But this apparent erasure of the subject is perhaps better understood as the enshrinement of Kant-style universal epistemic subject, whose standpoint simulates that God. Such a standpoint has allowed us to imagine and intervene in things at very high levels of abstraction and in ways that can only be justified in terms of the power unleashed by the resulting systematic view of things. The costs incurred have included devaluing our most immediate experiences of nature and subjecting things to quite artificial conditions for purposes of extracting knowledge.

For Francis Bacon and the other early Scientific Revolutionaries, this was a fair price to pay for doing divine work – God, after all, was thought to be himself transcendent and perhaps even alienated from nature (which helped to explain the apparent involved in the completion of creation). But without this theistic assumption, it becomes difficult to justify the unfettered pursuit of science, once both the costs and benefits are each given their due. Of course, we could simply say that science is what turns humans into gods. For all its hubris, this response would at least possess the virtues of candour and consistency. As it stands, scientists today shy away from any such strong self-understandings, preferring to hide behind more passive accounts of their activities – they

DOI: 10.1057/9781137277077

'describe' rather than 'generate' phenomena, they 'explain' rather than 'justify' nature, etc. Lost in this secular translation of an originally sacred mission is the scientist's sense of personal responsibility *qua* scientist. In the next and final section of this chapter, I explore what resuming such responsibility might mean in Humanity 2.0.

3.4　Faith and reason in Humanity 2.0: cybernetics as 'artificial theology'

The politically correct tendency of our day is to treat science and theology as 'separate but equal' life-support systems of thought. Yet, as I have argued, a more historically accurate rendition would see them as fundamentally inseparable. In the case of Western science and Christian theology, the relevant point of intersection has been *technology*, an observation that has been usually made in a spirit damning to both (e.g. Noble 1997). But we always need to keep in mind *that one person's sense of alienation is another's sense of self-transcendence*. Thus, I shall pursue a more synergistic sense in which technology has represented the intersection of Christian theology and Western science. I dub this *artificial theology*, understood as 'natural theology 2.0', a conception already implicit in Simon (1977), which *inter alia* considers natural selection as a design feature of divine creation that might inform the optimisation strategies we use to bring about our own projects.

My touchstone is the epigraph to my 2010 book, *Science: The Art of Living*. It comes from the conclusion of Norbert Wiener's 1950 classic popularisation of the cybernetic worldview, *The Human Use of Human Beings*: 'Science is a way of life that can flourish only when men are free to have faith.' Were this sentence uttered today, it would be most naturally read as implying that faith provides a normative basis for steering science in a humane direction. This in turn suggests a division of labour – what Stephen Jay Gould (1999) called 'non-overlapping magisteria' – whereby 'faith' provides a value orientation to science that it would otherwise lack, since scientific knowledge *ex hypothesi* is simply a means that may be used to serve many possible ends. However, Wiener meant almost the exact opposite. For him, the relevant sense of 'faith' is in science itself as a guide to the values that should inform the human condition, understood as something always in transit. Wiener presumed that science had taken up theology's task of defining who we are by posing successive challenges that involve distinguishing what is of universal concern and what is of only parochial interest.

DOI: 10.1057/9781137277077

We normally think about such challenges in broadly humanitarian terms – as demanding that we meet our common human needs before satisfying more specific desires which, were they to take precedence now, would most likely be deleterious to the interests of much of the world's population. However, Wiener was inviting us to paint on a broader onto-logical canvas. Might it not be the case that the very biological bodies of our birth belong on the side of the 'parochial' when it comes to charting the future evolution of humanity? In that case, our 'humanity' advances with our increasing versatility in channeling 'information' and 'energy' – which, if not outright identical terms, at least may be understood as referring to entities that are governed by the same thermodynamic principles. In the end we become expert optimisers. In other words, we are capable of making the most from the least. At the limit of such an imperative of efficiency is the Augustinian idea of divine agency as *creatio ex nihilo*: 'creation out of nothing'.

The 'we' in the preceding paragraph refers to that part of *Homo sapiens* that aspires to self-transcendence – a divine standpoint, which Wiener identified with the Greek *cybernos*, the pilot of a ship, whose Latin cognate is 'governor'. The bodily organ with which this function has been most clearly identified is, of course, the brain – which explains its centrality to the general education course I propose in the epilogue to this book. Behind the *cybernos* image is the idea that it takes only the right shifts of the wheel to steady a ship's course to a destination. These shifts are ideally regular and slight, in the manner of compensating for local error but could be sharp when needed, resulting in something more akin to a Gestalt switch. Assumed in all this is that the material world provides various forms of resistance to the realisation of our ends – an endless struggle against 'entropy', as Wiener would have normally recognised it from thermody-namics. However, in *The Human Use of Human Beings*, he also recognised it as a distinctly 'Augustinian' struggle. Following St Augustine, for Wiener, matter is not a separate evil force that seeks to undermine our own (let alone, God's) best-laid plans with death and decay. Rather, 'evil' simply personifies our own ignorance of what it takes to make the most of our own creativity, which is in turn the source of our divinity. In a nutshell: evil = inefficiency = disorganisation = a surplus of waste to value.

The ideal of efficiency is most recognisable in the history of Christianity from the positive connotation given to 'poverty', especially by the men-dicant friars – the followers of St Dominic and St Francis of Assisi – who provided the teaching staff for the first universities. The mendicants

DOI: 10.1057/9781137277077

(literally 'beggars') managed to thrive because they showed they could always make more of whatever little they were given. Their productivity was seen to emulate in human terms God's own creativity. The most obvious academic survival of this process occurs in the construction and delivery of the curriculum, whereby knowledge that was originally obtained by disparate and often quite painstaking means – that is, as research – is made available to students at a much lower cost in terms of the time and effort demanded. This in turn opens up opportunities for them to push back the frontiers of human ignorance still further.

Of course, in the last 500 years, this 'economisation' of inquiry has accelerated as our cognitive powers have become increasingly alienated and autonomised – or, to put it in terms of today's euphemisms, 'offloaded' and 'modularised', respectively. Specifically I have in mind the information and communication technologies that, in Donald Norman's (1988) sense, 'smarten' the human life-world by restructuring the physical environment so that it more readily yields sense. Were this general trend accorded the world-historic significance that it deserves, the US Transcendentalist Ralph Waldo Emerson would be deemed a great prophet of our times. Marshall McLuhan (1965) certainly thought so, as Emerson provided his inspiration for regarding 'media' (his term for information and communication technologies) as the 'extensions of man'. The original Emerson (1870) quote is worth recalling:

> The human body is the magazine of inventions, the patent office, where are the models from which every hint was taken. All the tools and engines on earth are only extensions of its limbs and senses.

Replace 'magazine of inventions' with 'Swiss army knife', and the quote could have come from the logical positivist Rudolf Carnap, who wanted to develop inductive logic as an instrument for measuring the progress of humanity's various self-extensions (Fuller 2007b: 79).

What we call the 'scientific worldview' is this trajectory taken to the limit. In the conduct of science, our cognitive instruments and objects are extended to standpoints and domains far removed from those familiar to our unadorned senses. (Think microscopes and telescopes.) The philosopher of science Paul Humphreys (2004) has gone so far as to argue that the 20th century revolutions in relativity theory and quantum mechanics present the prospect that the conduct of science is better suited to specialised data-gathering machines tied to computers than to human beings, if science is ultimately about obtaining the most comprehensive

DOI: 10.1057/9781137277077

and accurate account of physical reality. However, Humphreys' self-styled 'computational empiricism' may have given up on humans too quickly. There is another way to think about the matter: The more we are able to incorporate, say, distant stars into the same formulae and narratives used to explain earthbound events, the more we implicitly acknowledge that our normal bodily existence is simply a vehicle not an end in itself. In this respect, the modern philosophy of rationalism that harks back to Pythagoras and Plato but begins in earnest with Descartes and Leibniz – and of which Wiener was a latter-day descendant – is simply mysticism tethered to mathematics and machinery. The end of science remains that of mysticism: to simulate God's 'view from nowhere'.

Given this perspective, it is profoundly misleading to follow Max Weber's characterisation of science's pervasion of the life-world as 'disenchantment'. In fact, the only thing that has been disenchanted is the idea that our bodies and immediate surroundings provide the loci for meaning in the world. In the Protestant Reformation that idea – closely linked to Roman Catholic ritual – came to be seen as a fetish (or, in biblical terms, an 'idol') that failed to distinguish clearly our ultimate end from the various means that we might use to achieve it. The revolution in sentiment accompanying the Scientific Revolution of the 17th century meant that we acquired the capacity to identify with the entire universe in a way that made all of it of actionable concern (Koyre 1957). This was most immediately felt at the level of ethics and politics, with the rise of universalist principles of justice and governance, which presupposed standards applicable to people living in vastly different circumstances, on the basis of which aliens might rationally justify the radical reorganisation of the natives' living conditions. But such universalism also extended to what the modern period began to call the 'universe', namely, the totality of physical reality. In this respect, Newton's laws made good on the astrologers' dream that we might quantify the exact terms on which the universe hangs together as a whole for purposes of strategic adaptation and intervention. Latter-day manifestations of this perspective include the search for extraterrestrial intelligence and the call for human colonies in the cosmos as a solution to our earthbound environmental problems.

It might seem that by this point we have left the issue of faith far behind. Nevertheless, the relevant sense of 'faith' invoked in the Wiener epigraph is quite close to its etymological meaning as 'loyalty' or 'trust'. In particular, we need to have faith in the value of the massive self-transformation of our being that is compelled by the advancement

DOI: 10.1057/9781137277077

of science. To be sure, the process is far from seamless and subject to many setbacks, as the 20th century's two world wars demonstrated. In this respect, faith in science instills the same principle of perseverance that has underwritten the collectivist version of the Christian salvation narrative, in which deliverance from evil (or error) is seen to be achieved not in one's lifetime but by future generations, each of which will generate new problems as it solves old ones. But in the end, the cosmic balance sheet will show that the trail of blood was worth it (Passmore 1970).

Of course, by the time that end-state will have been reached, we will have evolved not only cognitively but also emotionally: Matters that originally seemed inherently evil – including death and destruction – will be seen as short-term costs to achieve longer term goals. In the late 17th century, the branch of philosophical theology known as 'theodicy' was invented to open discussion on this matter, which had been always implicit in Christianity's self-understanding. It would be fair to say that neither dogmatic nor critical theologians have had the stomach to pursue this line of inquiry very far: If the very idea of second-guessing God's motives did not strike them as blasphemous, then the implication that divine ends justify inhumane means rendered the deity singularly unattractive. Indeed, once Charles Darwin encountered this image of God in Reverend Thomas Malthus' explanation of human population patterns (which were endorsed by that great defender of nature's intelligent design, William Paley), he attributed Malthus' finding not to the subtleties of divine providence but the blindness of natural selection. My own view – which I believe was shared by Wiener and other cyberneticians – is that Darwin had simply lost his nerve in giving up on God. Clearly he had hoped for a deity much more sympathetic than the one Malthus had on offer.

Here it is worth returning to a point stressed in Kant's classical definition of what it means to be in a state of 'enlightenment': Humanity comes into adulthood, thereby making us no longer dependent on the paternalism of either the church or the crown. Sometimes this form of words is taken to imply an endorsement of atheism, but that fails to follow the logic of Kant's analogy. When we become adults, we do not disown our parents but adopt their point-of-view, which of course is not the child's point-of-view. This is the point that Darwin apparently failed to grasp in his understanding of what it means for science to bring us closer to God. Science enables us to see that facts of nature that had looked extremely harsh and arbitrary when taken as isolated events and reflect strategically informed acts worthy of emulation in our path to cognitive and emotional

DOI: 10.1057/9781137277077

maturity. To be sure, the history of eugenics always provides a cautionary tale to any automatic embrace of this proposal, but that does not deny that the proposal itself is constitutive of the human project.

We live in a time when such popular theologians as Karen Armstrong (2009) and popular atheists as Alain de Botton (2012) join together in defending the enduring value of religion as the ideal institution for generating experiences of awe, be it in nature as a whole or specific human artifacts like churches and hymns. In this context, religion is treated as an instrument designed to release us from self-centeredness so that we may live more harmoniously with our surroundings. Indeed, it would not be unfair to see both Armstrong and de Botton, in their own ways, trying to reverse the achievement of the Protestant Reformation as a springboard to modernity. Here both the theologian and the atheist seriously short-sell the role of God as an inspiration for human self-development, which inspired the rise of science in Christian Europe. In the modern period, the view that I have suggested here has been advanced by Unitarians, Deists and Transcendentalists – all religious dissenters notorious for their views of God as a somewhat chilly being who is detached from the day-to-day lives of his creatures. Wiener himself was the product of Unitarian education.

Let me conclude simply by noting that the sort of God adumbrated by these dissenters may already inhabit our minds as what the cognitive psychologist Daniel Kahneman (2011) has recently called 'System 1'. For Kahneman this is the aspect of our mental activity that thinks fast and intuitively, aiming to incorporate peak experiences into memorable wholes while removing matters of detail and duration that define the texture of subjective experience. On the one hand, a mind operating solely on System 1 would have be unreliable in matters of recall, detached from the specifics of one's own life and perhaps even prone to fantasy; on the other hand, System 1 empowers the mind to use the past as a basis to move confidently into the future. David Eagleman (2009) has drawn on his expertise to explore what might be called the 'neurotheological' implications of the idea that the brain is configurable as an entity that thinks in such a godlike manner. But complementing System 1 is System 2, which is the realm of subjective experience that moves more slowly and more affectively as each event is savoured individually. Kahneman provocatively argues that System 2, despite its undoubted value in enabling us to make sense of our immediate surroundings and infuses our sense of uniqueness, nevertheless struggles to survive in our brains. Yet, might this apparent liability not be count as evidence for the divine in us trying to escape its animal confines?

DOI: 10.1057/9781137277077

4

Ethics for Humanity 2.0

Abstract: *I propose the characteristic ethical posture of Humanity 2.0 to be moral entrepreneurship, which I dub 'the fine art of recycling evil into good'. Moral entrepreneurship works by taking advantage of situations given or constructed as crises. After briefly surveying the careers of three exemplars of the moral entrepreneur (Robert McNamara, George Soros and Jeffrey Sachs), I explore the motives of moral entrepreneurs in terms of their standing debt to society for having already caused unnecessary harm, but which also now equips him with the skill set needed to do significant good. Such a mindset involves imagining oneself a vehicle of divine will, which would be a scary proposition had it not been long presumed by Christians touched by Calvin. I conclude that moral entrepreneurship looks most palatable – and perhaps even attractive – if the world is 'reversible', in the sense that every crisis, however clumsily handled by the moral entrepreneur, causes people to distinguish more clearly the necessary from contingent features of their existence. This leads them to reconceptualise past damages as new opportunities to assert what really matters; hence, a 'superutilitarian' ethic that treats all suffering as less cost than investment in a greater sense of the good.*

Fuller, Steve. *Preparing for Life in Humanity 2.0.* Basingstoke: Palgrave Macmillan, 2013. DOI: 10.1057/9781137277077.

4.1   Introduction: 'never let a good crisis go to waste'

Interviewed during the 2008 US Presidential campaign, Rahm Emanuel, soon-to-be Barack Obama's Chief of Staff, quipped, 'You don't ever want a crisis to go to waste; it's an opportunity to do important things that you would otherwise avoid' (Zeleny and Calmes 2008). Jonah Goldberg immediately pounced on this quote, posting it on the website of the leading American conservative political magazine, *The National Review*, as confirming the most shameless tendencies of what he dubbed in his best-selling book, 'Liberal Fascism' (Goldberg 2007). In what follows I offer a sympathetic case for what Emanuel unwittingly revealed and Goldberg had decried – *moral entrepreneurship*, the fine art of recycling evil into good by taking advantage of situations given or constructed as crises. Moral entrepreneurship is the ultimate extension of the entrepreneurial spirit, whose peculiar excesses have always sat uneasily with *Homo oeconomicus* as constrained utility maximiser, an image that itself has come to be universalised (Jones and Spicer 2009: chap. 7). Thus, a task of this chapter is to reconcile the two images in terms of what by the end I call 'superutilitarianism'.

My argument proceeds as follows. In the rest of this section, I discuss the conditions that in the past century have enabled the rise of the moral entrepreneur. On the one hand, people have been afforded unprecedented opportunities – both through the market and the ballot box – to express a need for a change in themselves and their societies. On the other, aspiring moral entrepreneurs, very much like the comic book superheroes who also surfaced in this period, possessed a risky competence: Yes, they promised to provide direction but at a cost, often quite high, at least in the short term. In effect, the lure of the moral entrepreneur has directly turned on 'his' (and so far the exemplars have been almost entirely male) ability to render every decision a crisis for which he is then uniquely suited to resolve. Section two briefly surveys the careers of three exemplars of the moral entrepreneur. They share the same qualities as Schumpeter's paradigmatic entrepreneur, Henry Ford, except that they lacked a novel product detachable from themselves. Indeed, like the superhero, they have repeatedly had to renew the need for their services. Section three explores the reason for this in the context of the motives for moral entrepreneurship. In effect, moral entrepreneurs are already in society's debt by having already caused what might appear in retrospect as unnecessary harm. Section four explores the mindset

DOI: 10.1057/9781137277077

that allows the moral entrepreneur to understand *both* that 'he' has been the source of significant and seemingly gratuitous pain yet by virtue of that experience is now positioned to do significant good. That mindset involves imagining oneself as a vehicle of divine will, which would be a scary proposition had it not been presumed by Christians touched by Calvin. The fifth and final section proposes that moral entrepreneurship looks most palatable – and perhaps even attractive – if the world is 'reversible', in the sense that every crisis, however clumsily handled by the moral entrepreneur, causes people to distinguish more clearly the necessary from contingent features of their existence. This leads them to reconceptualise past damages as new opportunities to assert what really matters. The result is a 'superutilitarian' ethic that treats all suffering less as a cost than an investment in a greater sense of the good.

At the outset, Goldberg's response to Emanuel needs to be made a bit more explicit. For Goldberg, 'Liberal Fascism' is founded on an opportunity that democratic politics makes generally available, namely, that someone may claim the power to amplify and focus what most people already believe or want. This person thus presents himself as a means to realise the collective end. In a neglected work of classical sociology, Ferdinand Toennies (2003) argued that such an employment prospect was opened up in the early 20th century by the discovery/invention of 'public opinion' as something lying in wait to be monitored, diagnosed and treated. Indeed, it continued a tendency that had begun in the second half of the 19th century, namely, the proliferation of elections for public office. This creates a demand for political heroes in democracies, comparable to the capitalist market's demand for entrepreneurs (Schumpeter 1942: 256–7). Each such survey or election effectively manufactures a small 'crisis', issuing in a 'call for proposals' (aka nominating conventions) for individuals possessing the relevant salvific capacities to come forward.

The adjective 'salvific' is meant to draw attention to the theological undertow of this turn of events. When theologians in the Abrahamic traditions speak of 'God's handiwork', they are normally conflating two distinct features of divine agency: (1) that whatever God does bears his signature and (2) that whatever bears God's signature is finished. Moral entrepreneurship exists in the space between these two moments. Moral entrepreneurs make a mark without necessarily hitting the mark. In past cosmologies, that space would have been inhabited by demiurges and angels, but today's moral entrepreneurs do not sign in God's name

DOI: 10.1057/9781137277077

but their own. They are cut-rate Christs for a world that appreciates a bargain. We derive comfort from knowing that a human-friendly hand is guiding us through a crisis, even if we experience the ride as awkward or even destructive in the short term, since we believe that some overarching good is being served. Perhaps more importantly, moral entrepreneurs are forgivable because they do what we might have done at the time, had we sufficient courage (and resources). And so, just as we would wish ourselves a second chance, so too should we wish them.

In the fictional realm, this mentality spawned the comic book 'superhero', whose superannuation as a job category was wittily captured in the 1986–7 comic book series and 2009 film, *Watchmen*, in which an alternate Cold War America bans superheroes due to the potential of their unpredictably forceful *modus operandi* to do more harm than good in an already crisis-prone world. But of course, we do not live in such a world. The *Watchmen* world remains a curious fiction that subsists in an ontological purgatory that manages to demystify the superhero without providing something more lifelike in its place. Rather, we live in a world that has come to treat the superhero as a virtual reality that is capable of being actualised from time to time – and increasingly so, in a time when the state is no longer the monopoly arbiter of what counts as real for its inhabitants. In this world, sheer shows of will and purpose count more than what, if any, lasting good they achieve (Morrison 2011: chap. 13). Not inappropriately, then, the Watchman who turns out to be the villain is named 'Ozymandias', recalling Shelley's poem about the Egyptian king whose stone monuments have been laid to waste by the ravages of time.

We readily excuse, if not marvel at, those who can score stunning effects of any sort. At the very least such effects bear the signature of their author, which (we presume) must itself involve a special skill, given that (as postmodernists never tire of reminding us) the world is such an incredibly complex place, subject to many interacting forces, in which any action – however large – is likely to suffer from diffraction, resulting in an array of outcomes that swamps any sense of original intent. From this standpoint, superheroes appear remarkable, even if they do not quite hit the mark (and even cause some collateral damage), since their *tours de force* succeed in bringing the mark (and often themselves) more clearly into the view. Indeed, assuming that eventually the mark is hit, the initial harm will have prepared the way for a greater good – and perhaps even a clearer sense of what that good is, the 'sublime object of entrepreneurship' (Jones and Spicer 2009: chap. 3).

DOI: 10.1057/9781137277077

If US Defence Department Secretary Donald Rumsfeld's 'shock and awe' air strategy that launched the Iraq War in 2003 comes to be seen as the first swift, sharp blow towards achieving lasting peace in the Middle-East, then it may be counted as an instance of 'superheroism' in the above sense. But the key feature of moral entrepreneurship is that responsibility for such heroics is devolved to individuals who take it upon themselves to use their unique superpowers to try to right the wrongs of the world. These moral entrepreneurs are much closer to the letter of the comic superheroes than had been the aspiring Fascists of the 1930s on whom the comic characters were loosely based. After all, to be worthy of their titles, *il Duce* and *der Führer* require a nation-state to lead – a point that greatly exercised Toennies and his colleague Max Weber as possibly representing the shape of things to come in modern democratic politics (Baehr 2008). In contrast, comic superheroes always exist uneasily with established authorities, often working at cross-purposes to them, given the idiosyncratic nature of their powers, the exercise of which serves to unify a populace that was heretofore internally divided.

4.2   Who are the moral entrepreneurs?

To flesh out the ideal type of the moral entrepreneur, I shall briefly review the careers of three exemplars – Robert McNamara, George Soros and Jeffrey Sachs – who I first introduced as 'moral entrepreneurs' in *Humanity 2.0* (Fuller 2011: chap. 5). Despite representing successive generations that spanned the second half of the 20th century, their biographies share certain general features: They all came from secure middle-class backgrounds, distinguished themselves in elite academic settings and benefited from the institutional ties that all of this implies. At the same time, they seem to have been driven to escape being defined exclusively in these terms, which in turn has led them to treat their cultural capital more in the manner of a speculator than a landholder. In other words, these moral entrepreneurs risked their accumulated advantage in novel and provocative ways by reaching for long-term goals, while ignoring or minimising – at least to their own satisfaction – intermediary issues that might have deterred lesser mortals. Their superhero prototype is Batman, a radically sublimated version of Dracula's vampire-aristocrat, whose survival depended on sucking the blood of the young who cross his path, a classic Marxist image of the rent-seeking landholder

DOI: 10.1057/9781137277077

(Neocleous 2003). Thus, updated and transfigured, Batman, even while he extracts surplus by day as manager of his family's fortune, is drawn by night to fight characters such as the Joker who live parasitically off the social disorder indirectly generated by such wealth (Morrison 2011: chap. 1).

True to superhero form, the track records of our moral entrepreneurs are chequered, though they consistently made a strong impression. The persistence of these efforts appear to pay-off in the long run, as the qualities that contributed to their originally having caused much harm are later seen as having been necessary or even constitutive of what is now taken to be the good that has resulted from their actions. In all this our trio have been aided by a rather rigorous macro-economic perspective, albeit applied to rather heterogeneous targets in space and time. Their careers appear to confirm Joseph Schumpeter's thesis that the medieval origins of political economy were related to the theological disputes over *theodicy*, that is, the rationalisation of suffering as an expression of divine justice (cf. Schumpeter 1954: 108–9; Fuller 2010a: chap. 7).

1 *Robert McNamara (1916–2009):* The six decades McNamara spent in industry and public service is arguably the most striking testimony to the scope for power opened up by an MBA. His 'superpower' rested on the training in cost accounting and systems analysis he received at the Harvard Business School, as McNamara lacked content expertise in all of his top management posts. He first turned his hand to the US Army Air Forces Office of Statistical Control in the Pacific Theatre of the Second World War, where he improved the efficiency and effectiveness of bombing missions. Then he moved to the Ford Motor Company, where he revived the firm's fortunes by innovating car models simpler and safer than its competitors. As Secretary of Defence under Kennedy and Johnson (a post he held longer than anyone else), he introduced cost accounting to the Pentagon, which led him to overrule particular military requests during the Vietnam War because he did not deem them systemically cost effective, a judgement that he eventually made of the war itself, which led to his resignation and immediate placement as head of the World Bank. In that role he boosted international contributions by focusing the Bank's efforts on poverty reduction, funds for which were earmarked specifically to poor nations with policies to reduce population growth rates and income disparities, as well as maintaining and restoring the natural environment.

DOI: 10.1057/9781137277077

2 *George Soros (1930–)*: The most publicly visible, if not influential, speculator of our times, Soros fits the Batman mould of the brooding superhero who would ideally prefer to be left to do philosophy than right the wrongs of the world's financial markets. But ours is not an ideal world, and Soros's 'superpower' consists in applying the critical rationalism that Karl Popper taught him at the London School of Economics to the world of arbitrage. It was a natural fit, since the arbitrageur stays in business only by assuming that, for any commodity, half the market overvalues it and half undervalues it – which means that a profit can be made simply by falsifying both: that is, buying low and selling high. Of course, the question is when the two opposing errors are sufficiently discrepant to make the biggest killing. Soros' special talent lay in his ability to gauge such matters in that market of markets, currencies, where one is confronted with a heterogeneous array of value indicators for an entire nation. As the finance ministries in the UK and Thailand learned to their chagrin in the 1990s, Soros more than adequately rose to the challenge. Yet, rather than pocket all of the profit, Soros has invested much of it in causes designed to mitigate – if not avoid – the worst effects caused by people such as himself, which he believes is symptomatic of a world whose inhabitants are insufficiently self-critical to use their freedom wisely.

3 *Jeffrey Sachs (1954–)*: The youngest tenured full professor in the history of the Harvard Economics Department (aged 29), Sachs soon became director of Harvard's notorious Institute for International Development, through which academics persuaded inflation-ridden economies, especially in socialist countries, to undergo 'shock therapy' involving the removal of price and currency controls and most state subsidies for a short period of time until stable markets were established for most goods and services. Every country where this policy was applied – from Russia and the former Soviet Bloc to Latin America – resulted in major suffering by the population, the severity and longevity of which was determined by whether the country already had a tradition of rule of law and property rights to institutionalise the emerging markets. Sachs' innovation here was to turn capitalism itself into a global export, one bearing the trademark of Harvard University, itself a private corporation that was subsequently sued by some of the countries in question. Nowadays Sachs is still a global exporter of capitalism – but on a larger stage and by more fundamental means.

DOI: 10.1057/9781137277077

Thus, as special advisor to the Secretary-General of the United Nations, Sachs founded the Millennium Development Project, which centres on achieving ecologically sustainable economic growth by encouraging the rich nations to invest in the infrastructure of poor nations to render them more productive and hence more efficient contributors to the global economy.

What makes moral entrepreneurs so 'entrepreneurial' is that they alter not only our sense of which persons or actions are 'right' or 'wrong', 'good' or 'evil', but also what those very terms mean. This second-order transformation is what Joseph Schumpeter (1942), with a nod to Marx, originally called the 'creative destruction' of markets that is the entrepreneur's calling card. Thus, moral entrepreneurs skate close to, perhaps sometimes falling over, the edge of conventional morality in pursuit of an ethic that from the outside looks like a version of 'the end justifies the means'. Their *modus operandi* can be understood in terms of the following chain of reasoning. Many morally prohibited actions would be, even if permitted, extremely uneconomical, largely because their outcomes would be irreversible: We would never be able to recover what had been lost. Thus, the 'prohibitiveness' refers to the cost of satisfying a relatively circumscribed need/want vis-à-vis overall welfare. For example, indulging one person's crime of passion deprives society as a whole of one able-bodied, multipurposeful member. It is the sort of intuition that ecological activists would like to see extended from an individual human life to entire animal species and possibly nature as such, all presuming that an indefinitely expanded 'moral circle' is a mark of our moral progress (Singer 1981). But if the issues relating to 'economy' in this conventional sense are removed, do the actions still look so bad? The answer to this question is work cut out for the moral entrepreneur, who sees 'matters of principle' as symptomatic of an unimaginative approach to economy.

It is worth recalling that Schumpeter's best example of an entrepreneur was Henry Ford, the innovator of the mass produced automobile (McCraw 2007: 266). Much of the original 'principled' objections that motorcars ruined the experience of travel and despoiled the natural environment evaporated once they started to roll off the assembly line on a regular basis. Indeed, the following 50 years witnessed an evolution in automotive design that increasingly protected both the passenger and the environment, thereby addressing – albeit by reducing – the original misgivings in terms of efficiency issues (e.g. maximum miles per gallon

DOI: 10.1057/9781137277077

with minimum carbon and sulphur emissions). Moreover, Ford himself preemptively struck against union fears that a multitude of cheap labour would be needed to manufacture all those cars by offering steady workers the best wage package and welfare benefits available in America at the time. Little surprise, then, that *The New York Times* (glowingly) dubbed Ford 'The Mussolini of Highland Park' (Kaempffert 1928). Let us now consider in a bit more detail the two-step process implied here, by which an allegedly immoral invention is converted into a normal, if not necessary, extension of our selves.

4.3   How does moral entrepreneurship work?

The first step to moral entrepreneurship is to disentangle permissibility ('ought') from feasibility ('can') issues: It is always much easier to forbid something if it is difficult to do in the first place. Thus Ford's innovation involved removing purely physical barriers to the proliferation of motorcars, which previously had been manufactured as a boutique product whose scaling up appeared prohibitive in terms of both direct and indirect costs. But once the car was made easily available, it was difficult to prohibit it, even on moral grounds. The second step is the rational response that moralists make to this revision of the feasible: Applying the higher principle of 'two wrongs make a right' (aka 'negation of the negation' in Marxspeak), they capitalise on it, if only to pay to realise a version of the order they would like to see. This explains the political appetite for 'sin taxes' on alcohol and tobacco (and gasoline?) and, at least amongst those on the political left, for capital gains and inheritance taxes. If one cannot dispose of an evil, one might still be able to achieve good as a by-product of evil. Thus, the revenues collected by taxes associated with sin and exploitation are often earmarked for restoring health and the environment, if not offering outright incentives for the manufacture of more salutary products. Notice that in the process, the moralists' goalpost has been shifted from maintaining a static ideal of the good that increasingly goes against the tide of history to achieving a functionally equivalent state of goodness in a continuously renewable fashion.

In sum, a previously non-negotiable principle becomes economically manageable, as the moral entrepreneur incorporates potential opponents in ways that enable history to move in the desired direction. This

DOI: 10.1057/9781137277077

mentality provides the ethical backdrop against which to judge the work of the entrepreneur, from whose godlike perspective what potential investors and consumers treat as 'principles' are better seen as auxiliary constraints vis-à-vis longer term, larger scale goals that entrepreneurs pursue. Put less delicately: The recalcitrant feature of principles can be overcome once people shift from thinking in terms of what should be done to the *style* in which something should be done. Since the value of principles is supposed to be inherent to actions but independent of outcomes, it should be easy to shift the discussion to style, which also shares that feature. Good examples here are provided by campaigns for 'ethical consumption', which aim to absolve consumers of the most heinous forms of capitalist exploitation without questioning capitalism as such. In this light, entrepreneurship would appear tantamount to arrogant manipulation, had the Abrahamic religions not already endowed all humans with the capacity to operate in just this fashion – albeit in the full understanding that we are born as failed gods whose redemption is far from guaranteed. I shall return to this point below.

As for Schumpeter, he had decidedly mixed feelings about what the two-step process says about mindset of the entrepreneur. His generally pejorative use of the term 'utilitarian' epitomises his overall sentiment (Shionoya 1997: 188–9). More precisely, Schumpeter means that the entrepreneur sustains the dynamic character of capitalism by expanding the horizons of commercial society, rendering new things 'marketable', a term that had been coined by the Austrian hero of the 'marginalist revolution' in economics, Carl Menger (177). If commodities are naturally subject to diminishing marginal utility (i.e. nothing is desired indefinitely), then given free trade and increased productivity, it would seem to follow that all markets are eventually saturated, spelling the end of capitalism. It therefore becomes imperative for new commodities to be brought to market on a regular basis, which in postscarcity societies involves promoting products as representing values that consumers perhaps already implicitly hold but without having had assigned prices to them, let alone treated them as products in potential competition with others for their custom. To succeed in this tricky task of reengineering consumer psychology, the entrepreneur must be able to apply creatively Bentham's 'pushpin as good as poetry' principle to translate any value into a common monetary standard that both sides of an exchange can apply to their respective advantage. In this respect, the extent of entrepreneurship in a commercial society can be measured by the amount

DOI: 10.1057/9781137277077

that firms spend on advertising, the alchemy that enables people to see even the slightest change in product design as an invitation to inhabit a new way of being (McCraw 2007: 258).

To be sure, this is entrepreneurship at its most romantic, in which selling people what they do not need appears as the intellectual challenge of expanding their existential horizons. But our societies are still subject to real and imagined scarcities that, at least in principle, can be satisfied by normal market mechanisms. Thus, the entrepreneurial impulse must come from the would-be entrepreneurs themselves, who have their own reasons to 'never let a good crisis go to waste'. This motivation is, broadly speaking, grounded in some sort of debt that one feels needs to be repaid, ranging from unfulfilled personal aspiration or unredeemed status expectation to recovery from a fall-from-grace (Brenner 1990). The last, as we have seen, can be especially relevant to *moral* entrepreneurship. In any case, entrepreneurs must induce a value crisis. First, to secure the capital they need to market the new good, they must induce a value crisis in would-be investors by forcing them to make a 'bird in the hand, two in the bush' comparison, which if successful leaves the investors – at least in the short term – more vulnerable to any unexpected changes in the normal economy. Moreover, upon bringing their new products to market, entrepreneurs induce a second crisis, as consumers are now forced to think in terms of tradeoffs that they may not have previously considered. Thus, the decision to purchase a motorcar forces one to weigh a concern for the natural environment against a desire for fast and reliable personal transport.

4.4   Moral entrepreneurship in search of a standpoint it can call its own

The political psychologist and student of counterfactual reasoning, Philip Tetlock (2003), would locate the entrepreneur's special talent in the art of conjuring with 'taboo cognitions', a state of mind in which we weigh the relative merits of what is normally considered a sacred and a secular value, if not an outright profanity. But as the deliberator negotiates this risky terrain, there is an opportunity for higher-order value integration, very much in the spirit of Hegelian sublimation, or 'sublation', whereby a conflict of opposites is resolved into a more comprehensive whole. Arguably this is exactly what Newton achieved

DOI: 10.1057/9781137277077

in *Principia Mathematica*, when he unified the heretofore qualitatively distinct realms of terrestrial and celestial motions under a single set of mathematical principles (Fuller 2010c). In the century prior to Newton – the so-called Scientific Revolution – philosophers, scientists and theologians entertained many such taboo cognitions as they tried to determine how a world with so many regionally variant empirical features (as lovingly detailed by Aristotle and his followers) nevertheless is (if the Bible is taken literally) the product of intelligent design (Fuller 2010a: chap. 5). That this risky shift in value orientation may have been purchased at a high cost – even beyond the contravention of Church dogma – is captured in Max Weber's term *Entzauberung* ('disenchant- ment'), whereby modern people live with the capacity to explain from first physical principles events that cannot be so easily justified from a strictly moral point-of-view (Proctor 1991). In most general terms, moral entrepreneurship is about bringing this sense of modernity to light as a problem for which it then attempts to provide the solution.

Thus, modernity's nagging moral intuitions do not disappear but migrate and reassemble, as per the second stage of the two-step process of successful entrepreneurship outlined above. In this way, for example, modern biology turns the act of upholding the sanctity of life into exer- cises in establishing physical thresholds for antenatal life and determin- ing fair prices for interpersonal (and even interspecies) organ transfers. It means that theologians are forced to argue on the same grounds with medical scientists, health providers and policymakers. While this proc- ess is often called 'reduction', Schumpeter was more correct to see it as emblematic of a methodological collectivism, what he saw as utilitarian- ism's bias towards socialism (Shionoya 1997: 296). Implied in this idea is the ultimate of taboo cognitions, namely, the capacity to enter 'The Mind of God' from whose standpoint all differences are negotiable in terms of a single harmonious scheme that the deity is trying to realise. While this theological idea appears to take us far from the precincts of political economy, in fact it helps to explain how the entrepreneur inserts himself as an agent of completion who brings to light seemingly irreconcilable differences only then to reconcile them.

A relevant secular point of reference is Antonio Gramsci's famed self- understanding as 'a pessimist of the intellect but an optimist of the will'. Given our knowledge of the physical universe, indeed, the principle of entropy alone, the detached standpoint of the contemplative intellect – absent a divine saviour – might well conclude that this is the worst of all

DOI: 10.1057/9781137277077

possible worlds, as all semblance of order eventually disappears. However, those who feel personally empowered (i.e. possess the will) to reorganise what otherwise would dissolve into conflict and chaos can turn ours into the best of all possible worlds. One can see this intuition at work in the superhero's *modus operandi* always seeming to involve a burst of new energy into what heretofore had been a moribund system. The task would involve not restarting the world from scratch but rearranging the world as it stands to convert its proximate minuses into ultimate plusses. In short, contrary to its popular connotations, the optimist is very much a hard-headed realist who sees any immediate loss or benefit as simply means, if not raw material, in pursuit of some larger end. Indeed, in a Stockholm Syndrome way, the imprisoned Gramsci may have seen in himself a flattering job description for his captor Mussolini, a Nietzschean superman, perhaps the most popular prototype of the moral entrepreneur.

When Nietzsche dubbed humanity's successor species as the 'superman' (*Übermensch*), he chose the word carefully. In particular, supermen are *not* what Darwin's theory of evolution by natural selection might predict our successors to be. After all, the next great global catastrophe – say, the product of anthropogenic climate change – may not play to the human qualities that are most highly valued in today's world. To assume that wealthy, well-educated people are better prepared to survive the cull of natural selection may turn out to be wishful thinking born of excessive self-regard that ignores the incapacitation that the rich have unwittingly suffered from their long-term exploitation of the poor, which may have made them dependent on those whom they treat as disposable. Thus, a conspiracy of Darwin and Marx could easily undermine aspiring Nietzschean heroes, unless they are capable of extending specifically human qualities, often in the face of resistance from a world ultimately indifferent, if not hostile, to human values. Thus, as the word suggests, 'superman' references not simply the next species that happens to occupy our ecological niche, but a better class of human, an improvement on what is already on the scene (Sorgner 2009). This idea is much loved by transhumanists, who unsurprisingly are inclined to resort to superheroes to make their case for 'enhancing evolution', as exemplified by the comic strip Superman's flexed bicep that graces the cover of Harris (2007), a popular book on transhumanist bioethics. The superman is capable of taking hold of the course of events – if not the entire process of natural selection – and bending them to his will, so as not to let a good crisis go to waste.

DOI: 10.1057/9781137277077

But if the image of Superman is appropriate for the moral entrepreneur, then it follows that this person who is always on the lookout for a crisis must be quite extraordinary in many, perhaps not easily foreseen or fathomed ways. Once again, the theological backdrop to the concept comes into view, since it makes sense to envisage the superman as *ultimately* good, despite whatever harms he causes along the way, only if the continual improvement of humanity is thought to converge on some optimal and coherent state of being, what Eastern Orthodox Christianity still calls *apotheosis*. But in that case, some quite ordinary human virtues – prudence and compassion are good candidates – may come to be seen as throwbacks to our more imperfect days as they continue urging us to be more cautious or lenient in our dealings with others than is required. (Kantian ethics takes off from this point.) Perhaps then it is no accident that superheroes remove themselves from the scene soon after their deeds are done, since it is not clear that lesser mortals could live with them on normal basis. For if, as the Abrahamic religions maintain, we are created in the 'image and likeness' of God, and God is defined as the being who in every act exemplifies all the virtues, both infinitely and indivisibly (e.g. no tradeoffs between the deity's goodness and power), then a human with godlike aspirations can never treat any single human virtue as an end in itself. Accordingly, our divinity is to be found in the capacity to see beyond short-term gratification of all sorts – even the moral comforts that prudence and compassion provide us – in pursuit of a larger objective and its ultimate reward.

This point most clearly harks back to the fallen state of humanity, in which virtues that are unified only in God are disparately endowed in humans at any moment in their history. Redemption is then a journey towards reintegration (Fuller 2011: chap. 2). A familiar modern take on this sense of redemption is the Marxist drive to overcome humanity's self-alienation under capitalism through the unity of theory and practice. While this impulse is often cited as evidence of Marx's 'humanism', that conclusion is far from obvious. Shortly after the end of the Second World War, Leo Strauss and Alexandre Kojève entered into a profound correspondence – ostensibly about Plato's *Republic* – on this matter (Strauss 2000). Both realised – Strauss in horror but Kojève in hope – that a single person who optimally functioned as both philosopher and king (i.e. theorist and practitioner) would govern in a literally 'inhuman' (though we might now say 'transhuman') fashion, in light of which the Communist experiment may be considered to have been a 'success' of

DOI: 10.1057/9781137277077

sorts. This rather severe line of thought had came into its own during the Protestant Reformation, when Calvinism cast a long axiological shadow over the secular worldview, one which Max Weber famously associated with 'the spirit of capitalism' (Weber 1958: chap. 2). Thus, for Benjamin Franklin, the morally upright person recognises when the zealous pursuit of one virtue turns into a vicious vanity that inhibits the realisation of other virtues. To declare 'Honesty is the best policy' is to do no more than to acknowledge the virtue's instrumental value, a corollary of which is to avoid excessive honesty in situations where withholding some of what one believes might increase the commonweal. Indeed, Franklin employed the phrase 'economical with the truth' without irony to capture just this strategy (cf. Fuller 2009: chap. 4).

4.5   Conclusion: moral entrepreneurship and the ideal of a reversible world

The philosopher Avishai Margalit has observed that in biblical Hebrew, when humans 'forgive' each other's sins, they are effectively agreeing to assume the burden of that sin, whereas divine 'forgiveness' involves writing the sin off, as in 'to forgive is to forget' (Margalit 2003: 185–6). Moral entrepreneurship is about exploiting the former so as to promote the latter sense of 'forgive'. Thus, the moral entrepreneur is someone who both takes it upon himself to solve other people's problems (i.e. by assuming their debts) and indulges the trust of those whom he would save (i.e. by becoming indebted to them). The good done in the one case is offset by the risks and harms sustained in the other, all in the name of some sense of cosmic justice, or theodicy. In effect, moral entrepreneurship is in the business of proliferating creditor/debtor relations across the human condition, perhaps even 'financialising' humanity as such.

Here it is worth recalling that 'finance' derives from the same root as 'finalise', which may be understood as the process by which purposes are attached to things that might otherwise remain without purpose. To be sure, many, if not most, human activities are sufficiently purposeful that one knows when they should end – and whether that end should count as success or failure. However, those activities also tend to be self-consuming. But what about activities the purpose of which is to allow for still more purposeful activity? The modern economic conception of finance is born of this idea, which had a rocky start with the medieval scholastics who fretted that the proliferation of creditor/debtor relations would turn the creation of something out of nothing (*creatio ex nihilo*) into a business, since earning interest from a loan or profiting from an investment requires nothing more of the creditor than a capital transfer. Given God's monopoly hold over the injection of new energy into the world-system, such a proposition would be a sacrilege. Nevertheless, people came to accept that time itself, traditionally seen as that great layer to waste, may not only erode but also enhance our productive capacities. In particular, by transferring part of our capacity to act to others in the short term, we may effectively increase our capacity to act in the long term. I say 'may' because failure is always possible, but in the world of finance this possibility is also anticipated as 'risk', for which the debtor is liable to pay at a previously agreed rate (Goetzmann and Rouwenshorst 2005).

An important linguistic innovation of the 12th century that facilitated this risk-seeking sense of finance is associated with the great Bolognese legal glossator Azzone Soldanus who reified the time spent away from the creditor in the hands of the debtor as 'what is between' – *quod inter est* in Latin (aka *interest*) – to mean the additional value accrued by money in the time spent between the provision of a loan and its repayment. Before the coinage of 'interest', the idea that one should charge for lending money was seen as taking advantage of another person's misery (Langholm 1998). However, Soldanus turned the old concept on its head, focusing on the opportunity that the loan provides the poor person to improve himself. After all, if the borrower can repay the loan with interest in the allotted time, then he will probably have generated a profit in the interim which, having raised his standard of living, would demonstrate his worthiness to have received the loan. In short, interest turns the loan into a moral test of the borrower that is made possible through temporary self-restraint on the part of the lender. As Weber (1958) observed, this moralistic slant on the loan, which ideally eventuates in the improvement of both parties, came to be especially emphasised in Calvinist theology, in which the charging of interest functioned as a vehicle of moral instruction. A secular descendant of this strategy of binding two parties together so that they mutually benefit by cancelling out each other's excesses was used in the 18th century to justify the virtue of markets, which were envisaged as forcing producers to be more practical and consumers more discriminating than each might be otherwise (Rothschild 2001).

DOI: 10.1057/9781137277077

At the start of this chapter, I alluded to how moral entrepreneurship arose through crises generated by regular elections for public office and public opinion surveys. In these contexts, the moral entrepreneur is positioned as the one who can 'stop the rot', 'stem the tide' and 'restore order'. All of this presupposes that ours is a *reversible world* – that is, one in which no gain but also, more importantly, no loss need be permanent. With sufficient vigilance, error can be caught and corrected. Karl Popper (1957) famously asserted that reversibility is a precondition of the 'open society', whereby democracy is a genuine ally of liberty, as well as equality. However, he said this as part of an anti-utopian argument, which implied that policymakers should only act in ways that are reversible in the world *as it is*. He did not make the further – and probably, by his lights, utopian – argument that we should aim to increase the world's reversibility so as to maximise the 'openness' of the open society. Moral entrepreneurs purport to be agents of the open society in just this expanded sense. They aspire to not only prevent or mitigate bad consequences but also, and more interestingly, undo damage that has been already done. And where that is not feasible, the moral entrepreneur aims to provide compensation to convert the damage into an affordable cost, or (ideally) an outright asset. Here reversibility acquires the expanded meaning of 'commutability' that renders damage a tradable asset in a collectively recognised value system, rather than a grievance nurtured and hoarded so as to become the trademark of personal and cultural identity. The guilt, shame and resentment associated with misconduct and missed opportunities would be shorter lived, as revisionist histories and second chances become increasingly available. Indeed, 'No Regrets', recalling Edith Piaf's signature song, might provide the leitmotif for the reversibilist mentality.

Evidence for this line of thought may be found amongst legal philosophers who have sought to justify the extended sense of 'payment for damages' that has surfaced in recent judicial rulings (Ripstein 2007). They appear to be harbingers of a new sensibility, in which a criminal's relationship to the victim is becoming more like that of a venture capitalist engaged in a hostile takeover bid – in this case, of someone else's life. Once the crime has been committed, the criminal is obliged to demonstrate it can provide a genuine benefit to the victim and those parts of society associated with the victim and affected by the crime. Seen this way, in most cases a prison sentence would be inappropriate and arguably only serve to intensify the debt that has been incurred.

DOI: 10.1057/9781137277077

The conceptual link between this revised view of criminality and moral entrepreneurship is provided by an aspect of the history of capitalism long stressed by world-systems theorists, according to which trading routes and market towns in medieval Europe evolved from the coerced exchanges of pirates whose opportunity for enterprise had been made possible by the failure of imperial bureaucracies and manorial economies to ensure an adequate distribution of goods and services (Wallerstein 1991: 125–48). The pirates thus turned a deteriorating situation into crisis, in which they were portrayed as both hero and villain, but in response to which a more durable social order was built that accomplished the work of the pirates within a system that replaced arbitrary personal losses with structurally imposed sanctions, such as tolls, taxes and licences to trade. This is the context in which 'entrepreneur' first entered the French lexicon in the 15th century (Sørensen 2008).

It is in the spirit of the reversible world to see the difference between entrepreneurs and pirates from a purely systemic standpoint – that is, in terms of the net distribution of harms and goods brought about by their activity. In that case, the most salient difference between the two groups may simply boil down to whether the creditor's permission is initially sought before the debt is incurred by the entrepreneur (yes) or the pirate (no). From the standpoint of a reversibilist, *only* the order of events in setting up the exchange differs. And this difference continues to carry ethical import only as long as it is reasonable for people to regard certain aspects of their being as non-negotiable, regardless of the benefits they might receive by forfeiting them.

To be sure, the advance of capitalism, especially through its monetised system of exchange, has eroded the absoluteness of this intuition. Everything now appears to have its price – save perhaps one thing: the human body itself, which remains the site of 'dignity', despite increasing challenge even on this front (e.g. Pinker 2008). Capitalist claims to universal free-trade have traditionally rested on the physical inviolability of traders, which at the level of pure principle is unprecedented in exchange systems (Collins 1999: 206–7). Nevertheless this final barrier is very much subject to erosion in our time, as opportunities arise for people to identify themselves less closely with the bodies of their birth. A range of activities broadly associated with the 'commodification of life' fall under this category, including the legal extension of bioprospecting (i.e. the conversion of genetic material into intellectual property) beyond plants to animals and humans, increasingly respectable calls for an open

DOI: 10.1057/9781137277077

and comprehensive trade in organs, stem cells and other biomatter, as well as the promotion of psychic migration to 'second life' computer avatars and, more futuristically, durable and powerful silicon substrata (Fuller 2010b). The next step would be the emergence of moral entrepreneurs who not only make these developments readily available but also provide compelling incentives for people – especially poor ones – to take them up in the name of the ultimate exercise of freedom, exchanging part or all of their current selves for something greater with which they can self-identify (Fuller 2011: chap. 5).

The kind of reversibility suggested here, which may go so far as to 'reverse' one's own death through silicon resurrection or absorption into a corporate cyborg, appears to take us light years from Popper. In one sense that is clearly true. Popper's norm of reversible policymaking was guided by what he called 'negative utilitarianism', namely, the minimisation of human suffering by eliminating the conditions that produce immediately experienced pain. Popper thought that it was only in terms of this policy orientation, which he associated with social democratic poverty reduction schemes, that moral progress could be objectively measured. However, Popper's studious anti-utopianism may have led him to sell his own ideal short. After all, reversibility was not meant to be an end in itself but a value in service to the ideal of the open society, in which as many options would remain open for as many people as long as possible. Stating the ideal so explicitly reveals just how strongly it is opposed to the classic utopias, which were invariably conceived as static resting points for all of human history, and hence 'closed societies' in Popper's sense. Again the contrast in imagery is theologically inspired. The moral entrepreneur defends the open society ideal by emulating the deity as the source not of the ultimately tranquil paradise but of endlessly creative energy. However, it requires an ethic of 'superutilitarianism' that tolerates short-term pain for long-term gain, or as Nietzsche said in another context, 'What does not kill him makes him stronger'.

'Crisis' shares the same etymological root as 'critic', both implying 'judgement', understood concretely as an act of dividing what matters from what does not matter in a particular case, the wheat from the chaff. However, in the original Greek context, this act of 'decision' (literally, in Latin, the act of making the cut) was not itself the outcome of a decision. Rather, judges in classical Athenian court cases were chosen by lot. In modern democracies, the element of chance is introduced by giving the ruling party limited, if any, control over the scheduling of elections.

DOI: 10.1057/9781137277077

But perhaps capitalism has been the most efficient generator of crises, specifically as by-products of its unconditional commitment to greater productivity. In practice, this encourages the counter-inductive strategy of supposing that, regardless of how normal and successful an activity may be, there are always better ways of achieving the same ends, if not better versions of the same ends. But these invariably involve significant risk in the form of 'radical resource redistribution', a secular euphemism for the dark arts of theodicy. Thus, the moral entrepreneur requires, at various points, that people part with their money, their values and even their lives in their godlike quest to turn an unwanted crisis into an unwasted opportunity.

DOI: 10.1057/9781137277077

5

Epilogue: General Education for Humanity 2.0 – A Focus on the Brain

Abstract: *Humanity 2.0 provides an opportunity to put the brain at the heart of general education. The ultimate value of encouraging general study of changing knowledge and attitudes to the brain lies in the organ's long standing as a concrete site for entertaining human self-transcendence. Though of clear theological provenance, this function has taken on renewed significance in recent years, as marked by, on the one hand, an increased understanding of the continuities between human and animal brains; and on the other, the unprecedented prosthetic extension of human brains. Thus, there is an opportunity to provide a radical alternative to the traditional liberal arts understanding of the integrated human body as the natural seat of 'humanity'. In that spirit, the curriculum proposed here bears the centrifugally oriented title: 'The Brain in the West: From Divine Instrument to Transhuman Icon'. Much of the course is justified in terms of the emerging interdisciplinary field of 'neurohistory', which takes the specificity of the brain – both its hard-wiring and its plasticity – as an opportunity for reintegrating the natural and human sciences by promising a richer sense of what it means to 're-enact' the thoughts of the past.*

Fuller, Steve. *Preparing for Life in Humanity 2.0.*
Basingstoke: Palgrave Macmillan, 2013.
DOI: 10.1057/9781137277077.

5.1   Introduction: why the brain?

At first blush, the brain is not a promising focus for a course in general education or 'liberal arts' that a student might take before specialising in one subject. While it might be possible to get the 'two cultures' in agreement on such a course, it would probably be to oppose its implementation. To the humanist, such a concentration on the brain would appear to capitulate to biological reductionism, minimising if not trivialising 'the life of the mind'. To the scientist, the focus would seem to be premature if not overrated, given that most of cultural history has transpired without much substantive understanding of how the brain works – and even now our knowledge is limited and highly contestable. However, both responses miss the brain's historic distinctiveness as a human organ, namely, its role as an interface, a point of translation and mediation between, in the first instance, one's own mind and body, but more generally, the divine and the human. In this respect, the brain has always had a 'vehicular' character, which in recent years has inspired the psychologist Susan Blackmore (1999) to dub the brain a 'meme machine' in homage to Richard Dawkins' (1976) view of organisms as 'gene machines'. The brain's vehicularity has inspired thoughts that in order to increase the efficiency of meme transmission, the organ may be substantially enhanced and perhaps even replaced by something else – such as a mechanical android – that escapes the brain's biological moorings entirely.

Not surprisingly, then, the ultimate value of centring a general education course on the brain may lie in its long standing as a concrete site for entertaining human self-transcendence. Though of clear theological provenance, this idea has taken on renewed significance in these 'posthuman' times that are marked by, on the one hand, an increased understanding of the continuities between human and animal brains; and on the other, the unprecedented prosthetic extension of human brains (Fuller 2011). While pointing in somewhat different directions, these two developments take the locus of concern outside the traditional liberal-arts understanding of the integrated human body as the natural locus of 'humanity'; hence, the centrifugally oriented title of the proposed module: 'The Brain in the West: From Divine Instrument to Transhuman Icon'. The rest of this Epilogue is divided into two parts, the first outlining the module, the second providing some motivating intellectual considerations for giving the brain such overall prominence. Much of my justification turns on the emerging interdisciplinary field of *neurohistory*, which takes the

DOI: 10.1057/9781137277077

specificity of the brain – both its hard-wiring and its plasticity – as an opportunity for reintegrating the natural and human sciences by promising a richer sense of what it means to 'reenact' the thoughts of the past (Smail 2008). The focus here is not only on the brain's inherited capacities but also, and perhaps more importantly, what people have done to and with their brains. In effect, the construction of internal and external psychotropic environments updates what 'liberal arts' means in Humanity 2.0.

5.2 The proposed module: 'the brain in the west: from divine instrument to transhuman icon'

This ten-week module aims to survey the history of Western thought from the standpoint of the brain, a locus of increasing interdisciplinary interest in the early 21st century. The evolution of our understanding of this organ has charted humanity's changing relationship to the divine, the natural and the social. The topics, readings and assignments are designed to be of interest to those specialising in the humanities, social and natural sciences, as well as the medical and information technology professions.

The key objectives of this module are as follows:

▶ An appreciation of the centrality of the brain as a site of not only contemporary scientific and policy interest but also of cross-disciplinary understanding – a clear case of blind men trying to make sense of an elephant.

▶ A grasp of the sociological contexts in which conceptions of the brain have been implicated, especially in terms of defining the evolutionary limits of humanity.

▶ A reciprocal grasp of how various planned and unplanned developments in human history have potentially altered the character of the brain, including the relationship to its possessor.

▶ An awareness of the relatively seamless way in which classic questions from theology and philosophy have been translated into the modern scientific discourses of medicine, psychology and neuroscience.

Two books, McGilchrist (2009) and Taylor (2004), are readily available and may serve as general purpose reference books for the module. Both

DOI: 10.1057/9781137277077

provide cross-disciplinary overviews of the history of Western enquiries into the nature of the brain. Though their authors are trained in contemporary neuroscience, their outlooks differ, with McGilchrist more humanistic and positive in outlook, and Taylor more social-scientific and critical.

Students are required to watch at least one of the following five classic films of the past 50 years – all cheaply available on DVD – in which the brain figures prominently in the technologies of social control. The assignment may take one of two forms: (a) an academic critique of one or more aspects of the film in light of issues raised in the module; (b) a dramatic script based on one or more aspects of the film in light of issues raised in the module. The five films and their general relevance to the brain are as follows:

- ▶ *The Manchurian Candidate* (1962) – programming assassins
- ▶ *A Clockwork Orange* (1971) – rehabilitating delinquents
- ▶ *Minority Report* (2002) – anticipating crime
- ▶ *The Eternal Sunshine of the Spotless Mind* (2004) – erasing memories
- ▶ *Inception* (2010) – implanting thoughts

Weekly Module Content

1 *Introducing the Cult of the Brain:* Students are introduced to how and why the brain became the defining organ of the human condition from the 17th to the 19th centuries. This is the path charted from Descartes's fixation on the pineal gland at the brain's base, through Swedenborg's focus on the cerebral cortex, to finally the Freudian view of the self as an extension of the entire nervous system (Gross 1998). Originally treated as the meeting point between our animal and divine natures, the brain came to be regarded by an increasingly atheistic science as a secular fetish, a view which, if anything, contemporary neuroscience has helped to revive (Hecht 2003).

2 *The Brain's Access to God:* Even before the advent of modern neuroscience, the brain was seen as the organ necessary to tap into our divine natures (even if the heart or liver were regarded as more crucial for normal life functions). Two models of such tapping – both prominent in Platonic Christianity – were the 'arts of memory' and the 'course of study', the former drawing on our divinely inspired creativity (Yates 1966), the latter on the prospect of our reabsorption into God's mind (Bonaventure 1993). Together they formed

DOI: 10.1057/9781137277077

the basis for the scientific method in the 17th century. For Giordano Bruno, whose demise Galileo studiously avoided, it meant a form of self-discipline that would enable us to acquire a state of knowledge *sub specie aeternitate*, which is to say, as if we stood 'above' the empirical world in both spatial and temporal terms – that is, the standpoint from which physics still operates.

3 *The Brain Merges with God:* The analytic philosopher Thomas Nagel (1986) popularised the expression 'view from nowhere' to capture the divine standpoint that Newton arguably rendered humanly accessible, which brought into focus the nature of our capacity for 'second-order thought', that is, to see the world as if standing outside of it. The modern history of this prospect begins in the 17th century with the theologically suspect field that Leibniz christened in 1710 as 'theodicy', which invited systematic speculation on God's motives, given the sub-optimal (or 'evil') nature of the world as it normally appears to our finite minds (Neiman 2002; Nadler 2008). It later developed into the logical and computational puzzles associated with cybernetic models of the brain (Wiener 1950, 1964) and has been subsequently embraced by those who would project intelligent design on a complex nature (Fuller 2010a, 2011).

4 *The Brain's External Relations to Mind:* The 17th and 18th centuries witnessed the modern formulation of the two main problems of the philosophy of mind, both of which highlight the role of the brain as a translation device: How is mind related to body, and how do minds relate to each other? The former problem was defined in terms of, so to speak 'foreign exchange' (*commercium mentis et corporis*), that is, how much of the right sort of matter is needed for the expression of intelligence (Funkenstein 1986). The latter was addressed by finding a basis for calibrating human similarity. Was it our common descent from God, as Descartes thought, or our common life situations, as Adam Smith held? Latter day versions are, respectively, Chomsky (1966) and Hayek (1952).

5 *The Brain's Internal Relations as Mind:* The 18th and 19th centuries saw the rise of 'associationism' as an account of thought based on neural conductivity, which seeded many ideas about the role of contiguity and similarity in establishing mental patterns, not least that higher (aka divinely oriented) forms of thought are produced by the 'synthesis' or 'integration' of nervous energy. Thus, for the dissenting Christian scientist-ministers who started this movement, most notably David Hartley and Joseph Priestley, the defining human attribute of 'free will' became the capacity to determine

DOI: 10.1057/9781137277077

the brain's focus of 'attention'. However, after Darwin, associationism came to be increasingly identified with behavioural conditioning in a sense that was neutral on matters of the brain but presumed the existence of innate associative tendencies, or 'operants', that could be externally manipulated (Passmore 1970).

6 *Composing the Brain:* The two main modern views of the brain's organisation emerged in the 19th century as offshoots of medical enquiries: on the one hand, a 'modular' view that envisaged the organ as subject to micro-version of the social division of labour (Desmond 1992); on the other, a 'holistic' view that depicted the brain as an especially sensitive self-organising form of matter (Star 1989). The former tended to favour more-direct, the latter less-direct medical interventions, which in turn served to bifurcate the history of psychiatry in the 20th century (McGilchrist 2009).

7 *Combining Brains:* Alongside paradigms for making sense of the individual brain were paradigms for characterising the common or collective features of brains, especially as they adapt to changing historical circumstances. Prior to any clear empirical accounts of genetic transmission, theories of 'common sense' and 'collective memory' existed as alternative accounts of the acquisition and transmission of our humanity that over the 19th century came to be known as 'culture' (Valsiner and van der Veer 2000). The recent growth of evolutionary psychology and 'neurohistory' provides a new context for exploring how mass exposure to psychotropic elements in the environment (e.g. diet) have rewired human brains, resulting in new forms of sociality and self-expression (Smail 2008).

8 *The Global Brain:* As the cult of the brain peaked in the 20th century, the evolutionary prospect of a 'world brain' loomed on the horizon, understood either as a single unitary entity or a parallel distributed process. The unitary world brain was described in sacred (Teilhard 1959) and secular (Wells 1938) terms, both stressing tendencies towards amplification and standardisation in global communications. The distributed world brain began as an update of the classic geopolitical 'balance of power' (Deutsch 1963), though Wikimedia nowadays offers a postmodern take based on the endless differentiation and democratisation of knowledge production (Keen 2007).

9 *The Mass Mediated Brain:* From the printing press to the internet, the mass media have functioned – both intentionally and unintentionally – to reform the brain's powers, leaving the organ better equipped to adapt to the ever expanding mental ecology in which it has been embedded. This matter

DOI: 10.1057/9781137277077

may be seen in positive or negative terms, as well as approached from the brain or the media side, resulting in four prospects: brain-positive (Dehaene 2009), brain-negative (Greenfield 2003), media-positive (McLuhan 1965), media-negative (Lanier 2010).

10 *Conclusion – Brains Shaped, Washed and Sold:* The final week reflects on the extent and import of humanity's attempts to control its brains. The history of this ambition recapitulates the Reformation ('evangelism') and Counter-Reformation ('propaganda') roots of modern 'brainwashing', which may be seen as, respectively, a cathartic purge of unsavoury associations or a reinforcement of latent virtues. The introduction of brain scanning in the field of 'neuromarketing' represents a more invasive and personalised development along this trajectory (Taylor 2004).

5.3 Motivating the module: Western intellectual history as told from the brain's point of view

While no one disputes the brain's singular role in defining the human condition, it is remarkable just what little difference that fact has made so far to the human sciences. Most accounts of history that call themselves 'materialist' do not explicitly deal with the brain as a topic, let alone an agent, of history. Even the field called *neurophilosophy*, now a quarter-century old, has mostly served as a platform for applying broad-gauged arguments from evolutionary psychology to address the nature of the mind (Churchland 1986). Here the brain's neural circuitry is simply the pretext for playing out a rather detailed Darwinian version of naturalised epistemology. In contrast, Daniel Smail's *On Deep History and the Brain* (2008), whose proposal for a discipline of *neurohistory* forms the bulk of this section, is a breath of fresh air in its discriminating approach to both the human and the natural sciences.

Neurohistory is an instance of what I have called, in another context, 'postmodern positivism' (Fuller 2006b: chap. 4), as opposed to the more familiar 'postmodern humanism' exemplified in historiography by Hayden White's (1972) landmark work *Metahistory*. Whereas White was focused on the human capacity to create a variety of narratives that project realities whose rhetorical force is independent of any correspondence to an original reality, Smail is more interested in blurring boundaries that had provided occasions for historians to mark origins, such as animal/human,

DOI: 10.1057/9781137277077

nature/culture and context/text. Thus, Smail is much taken by Darwin's metaphor of the geological record as a changing dialect for a language we have yet to decipher, which he then updates with the idea that human behaviour is the archive of the performances that enable the historian to infer how our genetic competence is worked out in particular environments. As suggested by this example, Smail appropriates recent developments in the bioevolutionary sciences with considerable discrimination. After castigating the lingering reluctance of his fellow historians to admit that human genetic potential has not changed substantially for at least the last 40,000 years, he then concedes that our brains remain quite susceptible to changes in the physical environment, not least through what is ingested into our bodies as 'drugs', which alter our normal physiological functions in ways that have potency in new environments.

On the one hand, Smail clearly believes that historians should operate more 'archaeologically' in a broad sense that ranges from acknowledging the contemporary relevance of evidence drawn from the distant past, including our ape ancestors, to interpreting ordinary historical documents as an archaeologist might, namely, as symptoms of material processes or conditions largely outside their producers' direct control. However one ultimately judges Smail's project, he deserves credit for clearly seeing the line of influence here that runs from Darwin through Nietzsche and Freud to Foucault. On the other hand, Smail also denies the excessive claims of evolutionary psychology, which alleges that our inherited 'primate' or even 'reptilian' brains somehow continue to exert a stranglehold on what we can now do, think and be. Such 'Ultra-Darwinism' is no less than a secular version of the Original Sin – yet without the prospect of salvation at the end of time. But more to the point, loose talk about our animal brains is empirically unfounded. In situations where we happen to respond to our environments as apes do to theirs, it is simply because the environments do not distinguish the genetic potential of humans and apes.

Of course, what it means to distinguish humans and apes after Darwin is not so straightforward, but Smail grasps the problem by making creative use of Richard Dawkins' (1982) thesis of the 'extended phenotype', namely, the capacity of organisms to reorganise their physical environments so as to increase their selective advantage. In humans this is what culture looks like from a biological standpoint. For a Darwinist, it is also a good way to simulate Lamarckian effects. Thus, instead of imagining that one generation genetically transmits its experience to the next, the

DOI: 10.1057/9781137277077

earlier generation is seen as having hit upon a way of restructuring its environment that serves to elicit some previously unexpressed genetic potential, which in turn provides selective advantage to the later generations for whom this new environment constitutes their life-world.

The usual way of thinking about how phenotypes get extended recalls B.F. Skinner's operant conditioning experiments, whereby if you alter an organism's reinforcement schedule, you alter its behaviour pattern. The key difference is that whereas the Skinnerian experimenter deliberately aims to alter behaviour, Darwinian natural selection is notoriously blind to consequences. Smail's postmodern – but no less scientific – twist on this paradigm incorporates a very broad sense of 'psychotropic mechanisms' as part of humanity's extended phenotype. They go far beyond the architecture of the natural ecology. Indeed, Smail appears to believe that our brains and our bodies, what the founder of experimental medicine Claude Bernard originally called the '*milieu interieur*', have undergone the most significant reconstruction in human history. Humans, taken both individually and interactively, have come to function in ways that radically break with our simian past, even though our basic genetic makeup has not substantially changed in the interim.

Among the sources of this break are not only 'drugs' ingested as food and drink but also forms of sacred and secular discipline that together have served to render human culture portable. While it is an anthropological commonplace to say that humans became civilised only once they settled down to the largely sedentary existence that divides farmers from hunter-gatherers, equally – if not more – important for sharpening our distinctiveness from the apes has been the discovery of means for producing these civilising effects more or less on demand, regardless of location. The vulgarised version of this insight is that we might someday be able to take a pill – or implant a silicon chip – in order to think, do or feel as we please. Truth be told, the idea had begun to be floated once 'free will' was proposed as a distinctly human capacity. But its nature continues to elude evolutionary psychologists, who puzzle over the seemingly inexhaustible human faith in education, understood as a quest to achieve an ideal state of being rather than simply to avoid behaviour that results in harmful consequences. The latter, more modest and straightforwardly adaptive condition appears to satisfy our simian cousins. To his credit, Smail recognises this puzzlement on the part of evolutionary psychologists as reflecting their failure to grasp fully what it means to be human.

DOI: 10.1057/9781137277077

To put Smail's insight in evolutionary perspective, we might say that humans are not content simply to accept the adaptive value of particular traits in particular environments: In addition, we wish to generalise – if not universalise – some of those traits to new environments, even if *prima facie* they are not especially adaptive. In that case, attempts are made to redesign the new environments to enable the promotion of those traits so that they *become* adaptive. New texts, practices, buildings and comestibles are spawned. The spread of great 'civilisational' ideas such as Christianity, Islam, democracy, capitalism, socialism and science might be understood in just this way. However, as someone open to the idea of 'neurohistory', Smail is careful not to cast the spread of these ideas in excessively rationalistic terms. It is less a matter of people accepting the ideas based on external displays of logic or evidence than their feeling empowered in certain ways by how the ideas are embodied in their person. This last point may be especially important when it comes to the adoption of novel practices that cannot be yet justified on their own terms. Here Smail follows in the footsteps of Blaise Pascal and William James, who thought about belief in God in such terms, James (1902) having appreciated the neurochemical side of the matter.

For a non-religious example, consider Hirschmann's (1977) famous query of why it was that so many 17th and 18th century philosophers, theologians and politicians extolled the virtues of capitalism, even though it had delivered relatively little up to that point and was bound in the long term to destabilise long-standing feudal economic relations. Hirschmann's answer can be easily understood in neurohistorical terms: The practises associated with capitalism functioned therapeutically to tame the passions that had been held responsible for the political and religious unrest that marked the early modern period. People started to regard what transpired in their own minds, between themselves and their fellows, as well as the world at large, in more detached and calculating terms. Of course, as Smail himself stresses, such a general shift in a society's emotional landscape – its 'collective mood' – invariably has all sorts of unintended consequences, as people in that frame of mind are forced to respond to particular situations in various times and places. The overall result may be to convert a sense of self-discipline into an obsessive-compulsive disorder, which is certainly one way of making sense of what Marx later identified as capitalism's tendency towards a falling rate of profit.

DOI: 10.1057/9781137277077

Smail promises to breathe new life into the historical study of what in the 20th century would have been called 'worldviews', 'mentalities', 'thought styles', 'collective psychologies' or even sometimes (*à la* anthropologist Alfred Kroeber) 'superorganisms'. Neurohistory promises to confer a fluidity on these entities that breaks with their original image as monolithic mindsets restricted to a specific time and place. Smail's attempts at reenactment point to a day when a historian might undertake the combination of learning, discipline, diet and experience needed to approximate in her own person the mental dispositions of the historical agent that she wishes to understand: historiography as a form of 'method acting', if you will. To be sure, there have been precedents – starting with the philosopher of history Robin Collingwood's own reenactments of Roman life in northern England, and including Lewis Leakey's making of stone tools to understand the Palaeolithic mind and E.R. Dodds' climb of Mount Citheron to better grasp Euripides. But Smail's much more thoroughgoing reform of the historical method aims for no less than a tearing down of the Berlin Wall that continues to divide the *Geistes-* from the *Naturwissenschaften*.

The reader may feel disoriented by the juxtaposition of neuro-science and that classic method of humanist historiography, cognitive reenactment. However, it is worth recalling that when in the late 19th century Wilhelm Dilthey proposed empathy as a vehicle for inhabiting past minds, he based its feasibility on humanity's common ancestry, even though he was in no position to specify empirically the relevant features of our shared nature that might underwrite empathy's efficacy (Harrington 2001). Clearly Dilthey had been drawing on the theory of hermeneutical understanding previously proposed by the theologian Friedrich Schleiermacher, whilst trying to shift the sense of 'common ancestry' from the spiritual (i.e. our having been created *in imago dei*) to something more biopsychological that might be underwritten by evolution. Indeed, at the same time that Dilthey was secularising our powers of what Schleiermacher originally called 'divinatory understanding' (aka the capacity to look into another's soul, what nowadays is often called 'mindreading'), he was also defending what turned out to be the losing side of the battle to locate the newly emerging science of 'biology' in the hierarchy of sciences. Whereas he wished to see a vitalist version of biology figure as the foundational science of the *Geisteswissenschaften*, the mechanistic vision that came to prominence – popularised by Emil

DOI: 10.1057/9781137277077

DuBois-Reymond – portrayed biology as a specific complex extension of the *Naturwissenschaften* (Veit-Brause 2001).

Moreover, Dilthey was hardly alone amongst late 19th and early 20th century social and cultural theorists who thought of the brain as the material basis for cognition. An idea that was especially popular in France that influenced early thinking in American sociology was that patterns of excitation and inhibition in the brain might provide clues to spontaneously formed patterns of affiliation that not only transcend historical distance but also undercut the class and status markers that were increasingly proving to be a source of social conflict. (Freud's concept of the unconscious partly arose from this context.) This fuelled the idea that properly wired, or even rewired, brains – say, through the power of suggestion – might provide an all-purpose solution to society's problems (Turner 2007). The recent alleged discovery of 'mirror neurons' (i.e. neurons in the frontal lobe of the cerebral cortex that fire in the same way when watching or performing an action) has revived this line of thought, though most of the original conceptual problems involved in drawing inferences from brain patterns to socially relevant behavioural patterns remain (Gerrans 2009). But notwithstanding the specific existence of mirror neurons, there is no denying the brain's capacity for projective identification, perhaps most strikingly to 20th century psychologists, in the case of film, where it has been long recognised that people can identify with on-screen characters, even when they are presented in a highly stylised fashion, as in the case of abstract cartoons – a finding easily updated for avatars in cyberspace (Heider and Simmel 1944; cf. Turkle 1984).

In the line of thought and research outlined in the previous paragraph, the brain is envisaged as an organ with the capacity to generate realms of meaning that extend beyond, if not leave behind, its bodily basis. While the brain's self-transcending powers are easily understood in the context of religious experience, they may apply even more strongly to science, especially if physics is the paradigm, which also involves the sort of 'out-of-body' experience that has been philosophically objectified as the 'view from nowhere' (Nagel 1986) but finds its natural home in the brain's dream states (Koestler 1959). However, the ultimate bioevolutionary value of such feats of self-transcendence is far from clear, especially upon considering that the pursuit of science has arguably made our survival as a species increasingly precarious (Fuller 2010a: chap. 1).

The rest of this section proceeds first by showing that the brain, while key to our humanity, stands ambiguously with respect to life

DOI: 10.1057/9781137277077

itself – especially when compared with another organ, the heart. I then proceed to show that this ambiguity may be related to the interface role that the brain has played, at least since the late Middle Ages, between our divine and animal natures. The transplantation and immortalisation of brains, the conversion of nature to brain food and then technology, as well as our preoccupations with the aether and extraterrestrial intelligence, all point to the privileged status of the brain. The section concludes with some speculations about how literacy may have enhanced the brain's God-like creative powers.

The brain is humanity's most distinctive biological organ. It is the seat of our mentality. That much is clear. But is the brain the seat of life itself? That is much less obvious. Consider the continuing controversy over whether death should be legally defined in terms of the cessation of function in the brain or the heart (Brante and Hallberg 1991). In such a liminal state, the brain typically ceases to function before the heart does, but only the latter is normally defined as clinical death. This would suggest the prospect of life without a brain, which is certainly the standpoint of anyone who attributes life to creatures lacking vertebrae. But there is more here than meets the eye. Brain-death proponents tend to be quite literalist about the body's various 'organs' as instruments, which explains their fixation on thresholds beyond which an organ – especially the brain – is no longer of use to its host body. Thus, they want to invest more on research and equipment designed to establish such thresholds with greater precision and accuracy, typically for purposes of transplanting still usable organs. In contrast, heart-death proponents uphold a more discretionary notion of the distinction between life and death, one that comports well with a traditional reading of the Hippocratic oath as enjoining physicians not to cause harm to patients. In this context, 'life' is often seen as a divine gift even if someone harnessed to a life-support system is in no position to acknowledge the gift. A balanced judgement thus needs to be made between the interests of the patient and those of others who would suffer a permanent loss from the patient's death.

The difference in the locus of affect surrounding brain-death and heart-death is striking. The loved ones of a patient in a so-called vegetative state (i.e. with only autonomic brain activity) may call for indefinite life support, based on a religious belief in the sanctity of life, while enthusiasts for cryonics, fearing an 'information-theoretic death', may call for the brain to be extracted relatively early in the dying process to be preserved for future resuscitation before its higher functions

DOI: 10.1057/9781137277077

have a chance to deteriorate (Merkle 1992). This difference serves as a reminder that possession of a mind and a life, while subject to significant overlap in terms of the entities involved, are not identical states of being. Secular philosophers like to encapsulate the situation by saying that life is necessary but not sufficient for mind. To be sure, this conforms to the Darwinian narrative currently *en vogue*, whereby mind is an emergent property of biological processes that could not have been predicted. But this does not explain the increasing significance that has been attached to the brain specifically in the modern period.

Here theologians, at least of the Abrahamic faiths, offer a clue. For them biology implies an acceptance of the finitude of material beings, such that if one wishes to express a continuous train of thought, or 'spirit', indefinitely, then it will need to be reproduced through successive biological containers – that is, generations of organisms – each of which strives to make the thought more manifest over time. This is true whether one means the divine *logos* or a human institution. What began in Neo-Platonic theology as 'the great chain of being' (Lovejoy 1936) led to a natural and political economy of 'needs' by the end of the 18th century, and by the first decade of the 19th century eventuated in an evolutionary vision roughly equivalent to the one that Jean-Baptiste Lamarck championed under the very name of 'biology' (Bowler 2005).

However, organism provides only one of the two great images of the modern secular worldview. The other is the machine, already popular in the Middle Ages, which suggests that God could create a universe that is programmed to operate regularly, indefinitely and autonomously – which is to say, without the need for a replacement or improvement of parts (Rosen 1999: chap. 17). In their rather different ways, the mechanical clock and the steam engine governor captured this ideal in the 18th century (Mayr 1986). It is easy to forget that the machine's allure rests on what it says about its human makers: That we are the only animal capable of producing machines is indicative of our having been created *in imago dei* and hence provides clues to the divine *modus operandi* (Noble 1997). The ideal's allure resulted in the search for the perpetual motion machine, which came to grief on the concept of thermodynamic entropy in the late 19th century (Rabinbach 1990). Nevertheless, the search was resumed in the second half of the 20th century – this time with information replacing energy – as cyberneticians inferred that whatever practices distinguish humans most from other animals are probably what brings us closer to God. In that case, apotheosis would come from designing a mechanical brain (Wiener 1964).

DOI: 10.1057/9781137277077

The interesting general feature of this trajectory from the standpoint of humanity's self-conception is that the heart tends to stand metonymically for our 'organic' nature, while the brain so stands for our 'mechanical' nature. For those inclined towards structuralist poetics (Culler 1975), we might project heart-based relationships 'syntagmatically', which is to say, in terms of thinking of ourselves in lateral, symbiotic associations with other similarly embodied creatures, be they human or animal. On the other hand, brain-based relationships have tended to be 'paradigmatic', based on formal similarities between ourselves and other creatures who may be quite differently embodied and even normally operate on a different plane of reality (e.g. God, angels, machines). In the syntagmatic case, everything about us is potentially relevant to the formation of relations with other equally valued beings; in the paradigmatic case, we become self-discriminating, valuing only those aspects of our being that match the properties of higher order beings. At the late medieval dawn of the university, this divide was captured by the difference in scholastic ideology between the syntagmatic Dominicans (e.g. Thomas Aquinas) who presided over Paris and the paradigmatic Franciscans (e.g. John Duns Scotus) who dominated Oxford (Fuller 2011: chap. 2).

To put the contrast in sharp focus, consider what 'attraction' means in the two cases. In the heart-based world, attraction is founded on feeling, whereby emotion is conceived as virtual touch, as in bonds of sympathy highlighted in Adam Smith's (1759) *Theory of Moral Sentiments*. In contrast, in the brain-based world, 'attraction' is elevated to a state of higher receptiveness; hence, this view's historic fixation on light, which brings to mind the upward look of plants in phototropism, as well as the older astrological use of 'sympathy' to mean the relationship between stellar dispositions and human fates (Crombie 1953). The impact of this metaphysical conceit continues to be felt in the belief that something profound about the human condition will be revealed by advances in physical cosmology, despite its main focus on times and places alien to the human condition. Starting with Smith's contemporary, the Swedish mining engineer-turned-'spirit seeker' Emanuel Swedenborg, the cerebral cortex of the brain has been centrally implicated in this process of universal self-transcendence (Gross 1998: chap. 3). The Swedenborgian impulse is served today in the Search for Extraterrestrial Intelligence (SETI), which abandons biologically based bonds altogether for a notional 'cosmopolis' whose members are joined in the ability to transmit and receive binary code through radio waves. The only biological organ of ours that can

DOI: 10.1057/9781137277077

engage with that sense of universal citizenship is the brain (Basalla 2006: chap. 9).

Moreover, it is not clear that a brain requires a life in the normal biological sense, even granting that the organ's development has been so far the product of biological evolution. Here the history of technology proves an interesting witness. Its overall narrative thrust could be described as an extended attempt to reconstitute the world to make it easier for the brain to handle, even at the expense of the rest of our bodies, including the emotional aspects of our being that rely on more than the frontal lobe of the cerebral cortex. Thus, videogame theorists argue that the elective affinity between brains and games provides, in the guise of entertainment, opportunities for developing the complex social skills needed for solving real-world problems (McGonigal 2011). Indeed, intensive recreational gaming may even induce the rigorous dreaming of the sort dramatised in the 2010 film *Inception*, whereby the unconscious is not a hazy pastiche of half-remembered events but levels of eclectic architecture that quotes in novel combinations various times and places of relevance to the dreamer, very much in the spirit of how postmodernism arose in the arts (Kolb 1990).

Perhaps unsurprisingly, some brain scientists have expressed concerns that the increasing number of hours spent by humans staring at computers for both work and leisure is producing a form of overstimulation that is skewing our grasp of reality (Greenfield 2003). Such fears about the pathological amplification of one aspect of our being (call it 'synecdochosis') echo those generated by the rise of mass industrial labour in the 19th century. Back then the disproportionate application of muscular force, rather than neural conductivity, was seen as providing the greater threat to our humanity – namely, in the form of the body's thermodynamic decline, or, as it became known by the end of the century, 'fatigue' (Rabinbach 1990). Of course, in today's brain-centred world, we fear less that the body will wear itself out than that it will atrophy from disuse, as the brain itself falls victim to the contrary malaise of *sensory overload*. This condition, nowadays medically associated with autism, was coined by the German philosophical anthropologist Arnold Gehlen (1988) as *Reizüberflutung*. Like many of his fellow Nazi sympathisers, including Oswald Spengler, Othmar Spann, Carl Schmitt and Martin Heidegger, Gehlen believed that the characteristically modern tendency to lurch from crisis to crisis reflected the brain's technology-led overdevelopment, as society is continually reconstituted in line with the latest innovation,

DOI: 10.1057/9781137277077

or fad, the overall effect of which is to set our species adrift from its natural moorings.

In stark contrast to these reactionary 'back to nature' calls is the branch of engineering known as *biomimetics*, which develops new technologies in the spirit of treating life forms as paradigms of human utilities (Benyus 1997). It represents how biological science would look to a brain that was equally well disposed to being maintained naturally and artificially, as long as it were allowed to exert its capacity for creative thought (Rosen 1999: chap. 7). In that case, the bombardier beetle, say, is less interesting because it is a co-habitant of our biosphere or even that it occupies a particular offshoot in a common 'tree of life' than that it provides a prototype for improving payload delivery (*ScienceDaily* 2008). The insect is thus literally valued as 'food for thought' that is cerebrally metabolised, resulting in a technology with humanly relevant import. (The example is selected provocatively, since the research was done by the leading UK scientific proponent of intelligent design theory, Andrew McIntosh, for whom biology simply *is* biomimetics, given humanity's divine entitlement to complete creation.)

Today's philosophical normalisation of the brain's view of the world goes back to Putnam (1982), on the basis of which the idea that we might be 'brains in a vat' is taken as less an argument for scepticism about the external world than the simple acknowledgement that we might inhabit a world different from the one we first thought in terms of our causal relationship to it. To be sure, we would be forced to reconsider exactly what our words refer to, but without concluding that those words had become meaningless. In this context, the computer represents a high watermark as a technology designed to simulate the brain's multiple functions (aka software), in spite of radical differences in their physical constitution (aka hardware). After all, even the most primitive computer resembles a brain in terms of having been designed to integrate various digitally coded inputs for purposes of producing a co-ordinated sensorimotor response. It was just this line of thought that motivated Cold War cyberneticians to try to leverage the computer's initial success in capturing the brain's electrical circuitry (as what is now called a 'parallel distributed process') into a model of global governance, thereby making good on H.G. Wells' (1938) idea of a 'world brain' (Pickering 2010). Moreover, as Putnam's 'brains in a vat' scenario illustrates, the brain shares the computer's peculiarity in being both the means by which we investigate the world and the model of the world that we are investigating (Berry 2011).

DOI: 10.1057/9781137277077

An underexplored trace for tapping into this 'I am my brain' sensibility is the history of the *aether* concept. The Greeks tied it to the fire element, which the Stoics then domesticated into the 'breath of life' (*pneuma*), the prototype for the divine *logos* in Christianity. All of this traded on the very pervasive and ancient idea that the sun is the source of all motion. However, over the centuries, the sun's purely luminiferous qualities were disaggregated from its more generally energetic ones, resulting in the autonomisation and privileging of thinking (as 'reflection') above all other forms of animation. The first head of Oxford University, the Franciscan Robert Grosseteste, marks a pivotal turn in this direction, providing the context in which the brain eventually superseded the lungs and the heart as our organic interface with the rest of reality (Crombie 1953). The Franciscan quest for 'divine illumination' metamorphosed into the secular idea of Enlightenment, largely through Newton's immediate intellectual sources, the 17th century Cambridge Platonists, Henry More and Ralph Cudworth, the latter normally credited with having first called the faculty of reflection 'consciousness' (Passmore 1953).

But the most decisive move perhaps came in the second half of the 18th century by Christian dissenters, notably the physician David Hartley, who started to mean by 'aether' the medium through which all of reality transpired – including both light and thought – whose material character was at once an enabler and a retardant of humanity's capacity to reconnect to the source of all being from our presumptively fallen state (Allen 1999). This view gave clear sailing to the idea that scientific inquiry is a process of hypothesis-testing that corrects itself over many trials, in which individual observations are presumed to be 'errors' hovering around a real mathematical principle (understood – at least by Hartley and fellow dissenters such as Joseph Priestley – as an expression of the divine *logos*), for which later hypotheses then overcompensate, only to be themselves subsequently overturned (Laudan 1981: chaps. 8, 14). Such a metaphysically comprehensive approach to aether, which persists as late as the 1911 edition of the *Encyclopaedia Britannica*, came to define the theory and practice of 19th century physics, largely through the work of James Clerk Maxwell, who himself drew on innovative interpretations and uses of statistics in psychology and the social sciences, via such concepts as 'normal distribution', 'just noticeable difference' and 'marginal utility' (Porter 1986; Heidelberger 2004). This mentality persisted in the 20th century in creative attempts to lower the 'noise-to-signal' ratio in communications (e.g. Wiener 1950).

DOI: 10.1057/9781137277077

Ironically, given our secular times, after having taken seriously for two centuries that the brain might be the seat of the mind/soul, we are may be finally getting clear about what it would mean to adopt the divine standpoint. Indeed, one neuroscientist has come up with an ingenious way of exploring the matter, which takes seriously the tricky task of translating 'secular' back into 'sacred' time. Thus, Eagleman (2009) envisages the brain as a multiply biased recorder of human experience that serves the brain-bearer in life but effectively scrambles the ultimate meaning of that experience. However, at the moment of brain-death, God decodes the brain so to reveal to the brain-bearer the meaning of what s/he has just experienced. It is presented as a narrative consisting of familiar events now placed in their proper order, which is to say, weighted by their proper significance. Here intimate knowledge of the brain's workings is necessary to define the nature of the divine task at hand – aka The Final Judgement – and its (deceased) human reception of its outcome.

But what is it that has enabled us to simulate the divine standpoint, even after God has ceased to be of moral and practical relevance to much, if not most, of humanity? The answer may lie in the spread of *literacy*. According to Dehaene (2009), it is a triumph of the brain's plasticity, what he characterises as 'cortical recycling', whereby sensory inputs normally processed in different parts of the cerebral cortex are merged into a common activity that then causes the relevant neurons to develop together. Thus, reading merges the auditory input of speech and the visual input of writing to produce a simulation – if not hologram – of a fellow rational being. Dehaene here draws inspiration from one of the original stylists of the Spanish language, Francisco de Quevedo, who said that reading allows us to 'listen to the dead with our eyes' (Dehaene 2009: 325). Those dead may be understood as not only making a statement but also conveying an intention, the statement of which may or may not convey the intention fully, and so, given the chance, this imagined entity could expand in many ways not expressed in the writing but which the brain-bearer could project as imaginary dialogues with the dead. However, our brains are not simply populated by ghosts. We still retain our native sensory-based brain functions, which enable us to pick up patterns in both spoken and written text. This, in turn, allows us to acquire genre- and style-based knowledge, the value of which was often underestimated by 20th century academic humanists, who increasingly treated the brute act of reading – the 'closer' the better – as the only appropriate way of dealing with human expression (Gorman 2011).

DOI: 10.1057/9781137277077

Yet, generally speaking, facility with mixing and matching genres and styles – the stuff of improvisation – has been the fount of our expressive originality, perhaps the moment when the brain comes into its own (Fuller 2009: chap. 4).

DOI: 10.1057/9781137277077

References

Agamben, G. (1998). *Homo Sacer: Sovereign Power and Bare Life*. Palo Alto California: Stanford University Press.

Aharon, I. and Bourgeois-Gironde, S. (2011). 'From Neuroeconomics to Genetics: The Intertemporal Choices Case as An Example'. In R. Ebstein, S. Shamay-Tsoory and S. H. Chew (eds), *From DNA to Social Cognition*. (pp. 233–44) London: Wiley-Blackwell.

Allen, R. (1999). *David Hartley on Human Nature*. Albany New York: SUNY Press.

Appleyard, B. (2011). *The Brain is Wider than the Sky*. London: Weidenfeld and Nicolson.

Armstrong, K. (2009*). The Case for God: What Religion Really Means* London: The Bodley Head.

Armstrong, R. (2012). *Living Architecture: How Synthetic Biology Can Remake Our Cities and Reshape Our Lives*. TED e-book.

Baehr, P. (2008). *Caesarism, Charisma and Fate: Historical Sources and Modern Resonances in the Work of Max Weber*. Piscataway New Jersey: Transaction.

Barben, D., Fisher, E., Selin, C. and Guston, D. (2008). 'Anticipatory Governance of Nanotechnology: Foresight, Engagement and Integration'. In E. Hackett et al. (eds), *Handbook of Science and Technology Studies*. (pp. 979–1000) Cambridge Massachusetts: MIT Press.

Basalla, G. (2006). *Civilized Life in the Universe*. Oxford: Oxford University Press.

Becker, C. (1932). *The Heavenly City of the Eighteenth-Century Philosophers*. New Haven Connecticut: Yale University Press.

Benedetti, J. (1982). *Stanislavski: An Introduction*. London: Methuen.

Benyus, J. (1997). *Biomimicry: Innovation Inspired by Nature*. New York: William Morrow & Company.

Berlin, I. (1958). *Two Concepts of Liberty*. Oxford: Oxford University Press.

Berry, D.M. (2011). *The Philosophy of Software*. London: Palgrave Macmillan.

Bertalanffy, L.V. (1950). 'An Outline of General System Theory'. *British Journal for the Philosophy of Science* 1: 134–65.

Blackmore, S. (1999). *The Meme Machine*. Oxford: Oxford University Press.

Blumenberg, H. (1983). *The Legitimacy of the Modern Age*. (Orig. 1966) Cambridge Massachusetts: MIT Press.

Bonaventure (1993). *The Journey of the Mind to God* (Orig. mid-13th c.) Indianapolis: Hackett.

Bonnicksen, A. (2009). *Chimeras, Hybrids and Interspecies Research: Politics and Policymaking*. Washington DC: Georgetown University Press.

Borghol, N., Suderman, M., McArdle, W., Racine, A., Hallett, M., Pembrey, M., Hertzman, C., Power, C. and Szyf, M. (2011). 'Associations with Early-Life Socio-Economic Position in Adult DNA Methylation'. *International Journal of Epidemiology*. doi: 10.1093/ije/dyr147.

Bourke, J. (2011). *What It Means to Be Human: Reflections from 1791 to the Present*. London: Virago.

Bowler, R. (2005). 'Sentient Nature and Human Economy'. *History of the Human Sciences* 19(1): 23–54.

Brante, T. and Hallberg, M. (1991). 'Brain or Heart? The Controversy over the Concept of Death'. *Social Studies of Science* 21: 389–413

Brenner, R. (1990). *Rivalry: In Business, Science, among Nations*. Cambridge UK: Cambridge University Press.

Briggle, A. (2010). *A Rich Bioethics*. South Bend Indiana: University of Notre Dame Press.

Cartwright, N. (1983). *How the Laws of Physics Lie*. Oxford: Oxford University Press.

Castells, M. (2009). *Communication Power*. Oxford: Oxford University Press.

Chalmers, D. (1996). *The Conscious Mind: In Search of a Fundamental Theory*. Oxford: Oxford University Press.

DOI: 10.1057/9781137277077

Chan, S., Zee Y.-K., Jayson, G. and Harris, J. (2011). ' "Risky" Research and Participants' Interests: The Ethics of Phase 2C Clinical Trials'. *Clinical Ethics* 6: 91–6.

Chatterjee, A. (2007). 'Cosmetic Neurology and Cosmetic Surgery: Parallels, Predictions and Challenges'. *Cambridge Quarterly of Healthcare Ethics*. 16: 129–37.

Chomsky, N. (1966). *Cartesian Linguistics: A Chapter in the History of Rationalist Thought*. New York: Harper and Row.

Churchland, P.S. (1986). *Neurophilosophy: Toward a Unified Science of the Mind/Brain*. Cambridge Massachusetts: MIT Press.

Cochrane, A. (2010). 'Undignified Bioethics'. *Bioethics* 24(5): 234–41.

Collins, R. (1999). *Macrohistory: Essays in Sociology of the Long Run*. Palo Alto California: Stanford University Press.

Crombie, A.C. (1953). *Robert Grosseteste and the Origins of Experimental Science*. Oxford: Clarendon Press.

Culler, J. (1975). *Structuralist Poetics*. Ithaca New York: Cornell University Press.

Dawkins, R. (1976). *The Selfish Gene*. Oxford: Oxford University Press.

Dawkins, R. (1982). *The Extended Phenotype*. Oxford: Oxford University Press.

De Botton, A. (2012). *Religion for Atheists*. London: Hamish Hamilton.

De Grey, A. (2007). *Ending Aging: The Rejuvenation Breakthroughs that Could Reverse Human Aging in Our Lifetime*. New York: St Martin's Press.

Dehaene, S. (2009). *Reading in the Brain*. New York: Viking.

Desmond, A. (1992). *The Politics of Evolution: Morphology, Medicine, and Reform in Radical London*. Chicago: University of Chicago Press.

Deutsch, K. (1963). *The Nerves of Government: Nationalism and Social Communication, Political Community at the International Level*. New York: Free Press.

Dickens, P. (2000). *Social Darwinism: Linking Evolutionary Thought to Social Theory*. Milton Keynes, UK: Open University Press.

Dobzhansky, T. (1967). *The Biology of Ultimate Concern*. New York: New American Library.

Eagleman, D. (2009). *Sum: Forty Tales from the Afterlives*. NewYork: Pantheon.

Elster, J. (1976). 'A Note on Hysteresis'. *Synthese* 33: 371–91.

Emerson, R.W. (1870). 'Works and Days' in *Society and Solitude*. Cambridge Massachusetts: Welch, Bigelow and Co.

DOI: 10.1057/9781137277077

Fukuyama, F. (2003). *Our Posthuman Future*. London: Picador.

Fuller, S. (1988). *Social Epistemology*. Bloomington Indiana: Indiana University Press.

Fuller, S. (2000a). *The Governance of Science*. Milton Keynes UK: Open University Press.

Fuller, S. (2000b). *Thomas Kuhn: A Philosophical History for Our Times*. Chicago: University of Chicago Press.

Fuller, S. (2006a). *The New Sociological Imagination*. London: Sage.

Fuller, S. (2006b). *The Philosophy of Science and Technology Studies*. London: Routledge.

Fuller, S. (2007a). *The Knowledge Book: Key Concepts in Philosophy, Science and Culture*. Durham UK: Acumen.

Fuller, S. (2007b) *New Frontiers in Science and Technology Studies*. Cambridge UK: Polity

Fuller, S. (2007c) *Science vs. Religion? Intelligent Design and the Problem of Evolution*. Cambridge UK: Polity.

Fuller, S. (2008a). 'The Normative Turn: Counterfactuals and a Philosophical Historiography of Science'. *Isis* 99: 576–84.

Fuller, S. (2008b). *Dissent over Descent: Intelligent Design's Challenge to Darwinism*. Cambridge UK: Icon.

Fuller, S. (2009). *The Sociology of Intellectual Life*. London: Sage.

Fuller, S. (2010a). *Science: The Art of Living*. Durham UK and Montreal: Acumen and McGill-Queens University Press.

Fuller, S. (2010b). 'Capitalism and Knowledge: The University between Commodification and Entrepreneurship'. In H. Radder (ed.) *The Commodification of Academic Research: Science and the Modern University*. (pp. 277–306) Pittsburgh: University of Pittsburgh Press.

Fuller, S. (2010b). 'Thinking the Unthinkable as a Radical Scientific Project'. *Critical Review* 22(4): 397–413.

Fuller, S. (2011). *Humanity 2.0: What It Means to Be Human Past, Present and Future*. London: Palgrave Macmillan.

Funkenstein, A. (1986). *Theology and the Scientific Imagination*. Princeton: Princeton University Press.

Galison, P. and Stump, D. (eds) (1996). *The Disunity of Science: Boundaries, Contexts, and Power*. Palo Alto California: Stanford University Press.

Gehlen, A. (1988). *Man: His Nature and Place in the World*. (Orig. 1940). New York: Columbia University Press.

Gerrans, P. (2009). 'Imitation and Theory of Mind'. In J. Caccioppo and G. Bernston (eds), *Handbook of Neuroscience for the Behavioural Sciences*. (pp. 905–22) Chicago: University of Chicago Press.

DOI: 10.1057/9781137277077

Giridharadas, A. (2010). 'Where a Cellphone Is Still Cutting Edge'. *The New York Times*. 11 April.

Goetzmann, W. and Rouwenhorst, K. (2005). 'Financial Innovations in History'. In W. Goetzmann and K. Rouwenhorst (eds) *The Origins of Value: The Financial Innovations that Created Modern Capital Markets*. (pp. 3–16). Oxford: Oxford University Press.

Goldberg, J. (2007). *Liberal Fascism: The Secret History of the Left from Mussolini to the Politics of Meaning*. Garden City New York: Doubleday.

Goodman, N. (1955). *Fact, Fiction and Forecast*. Cambridge Massachusetts: Harvard University Press.

Gorman, D. (2011). 'The Future of Literary Study'. *Modern Language Quarterly* 72(1): 1–18.

Gould, S.J. (1988). *Wonderful Life*. New York: Norton.

Gould, S.J. (1999). *Rocks of Ages*. New York: Norton.

Greenberg, G. (2011). 'Inside the Battle to Define Mental Illness'. *Wired*. January. http://www.wired.com/magazine/2010/12/ff_dsmv/. Accessed 31 August 2012.

Greenfield, S. (2003). *Tomorrow's People: How 21st Century Technology Is Changing the Way We Think and Feel*. London: Allen Lane.

Gross, C.G. (1998). *Brain, Vision, Memory: Tales in the History of Neuroscience*. Cambridge Massachusetts: MIT Press.

Haraway, D. (1991). *Simians, Cyborgs, Women*. London: Free Association Books.

Harrington, A. (2001). 'Dilthey, Empathy and Verstehen'. *European Journal of Social Theory* 4: 311–29.

Harris, J. (2007). *Enhancing Evolution*. Princeton: Princeton University Press.

Hayek, F. (1952). *The Sensory Order*. Chicago: University of Chicago Press.

Hecht, J.M. (2003). *The End of the Soul: Modernity, Atheism and Anthropology in France*. New York: Columbia University Press.

Heidelberger, M. (2004). *Nature from Within: Gustav Fechner and His Psychophysical Worldview*. Pittsburgh: University of Pittsburgh Press.

Heider, F. and Simmel, M. (1944). 'An Experimental Study of Apparent Behavior'. *The American Journal of Psychology* 57: 243–59.

Hirsch, F. (1976). *The Social Limits to Growth*. London: Routledge & Kegan Paul.

Hirschmann, A.O. (1977). *The Passions and the Interests*. Princeton: Princeton University Press.

Hope, J. (2008). *Biobazaar: The Open Source Revolution and Biotechnology*. Cambridge Massachusetts: Harvard University Press.

DOI: 10.1057/9781137277077

Horgan, J. (1996). *The End of Science*. Reading Massachusetts: Addison-Wesley.

Humphreys, P. (2004). *Extending Ourselves: Computational Science, Empiricism and the Scientific Method*. Oxford: Oxford University Press.

Huxley, J. (1953). *Evolution in Action*. New York: Harper and Row.

James, W. (1902). *The Varieties of Religious Experience*. New York: Longmans, Green and Co.

Jones, C. and Spicer, A. (2009). *Unmasking the Entrepreneur*. Cheltenham UK: Edward Elgar.

Kaempffert, W. (1928). 'The Mussolini of Highland Park'. *The New York Times*. 8 January.

Kahneman, D. (2011). *Thinking, Fast and Slow*. London: Allen Lane.

Keen, A. (2007). *The Cult of the Amateur: How Today's Internet Is Killing Our Culture*. Garden City New York: Doubleday.

Kitcher, P. (2001). *Science, Truth and Democracy*. Oxford: Oxford University Press.

Koerner, L. (1999). *Linnaeus: Nature and Nation*. Cambridge Massachusetts: Harvard University Press.

Koestler, A. (1959). *The Sleepwalkers: A History of Man's Changing Vision of the Universe*. New York: Macmillan

Kolb, D. (1990). *Postmodern Sophistications*. Chicago: University of Chicago Press.

Koyre, A. (1957). *From the Closed World to the Infinite Universe*. Baltimore: Johns Hopkins University Press.

Krieger, L. (1957). *The German Idea of Freedom*. Chicago: University of Chicago Press.

Kuhn, T.S. (1970). *The Structure of Scientific Revolutions*. 2nd edn. (Orig. 1962). Chicago: University of Chicago Press.

Kurzweil, R. (1999). *The Age of Spiritual Machines*. New York: Random House.

Kurzweil, R. (2005). *The Singularity is Near: When Humans Transcend Biology*. New York: Viking.

Langholm, O. (1998). *The Economic Legacy of Scholastic Thought: Antecedents of Choice and Power*. Cambridge UK: Cambridge University Press.

Langlitz, N. (2010). 'The Persistence of the Subjective in Neuropsychopharmacology. Observations of Contemporary Hallucinogen Research'. *History of the Human Sciences*. 23(1): 37–57.

Lanier, J. (2010). *You Are Not a Gadget*. London: Allen Lane.

DOI: 10.1057/9781137277077

Latour, B. (1993). *We Have Never Been Modern.* Cambridge Massachusetts: Harvard University Press.

Laudan, L. (1981). *Science and Hypothesis.* Dordrecht: D. Reidel.

Lenski, R., Ofria, C., Pennock, R. and Adami, C. (2003) 'The Evolutionary Origin of Complex Features'. *Nature* 423: 139–44.

Longino, H. (2001). *The Fate of Knowledge.* Princeton: Princeton University Press.

Lovejoy, A.O. (1936). *The Great Chain of Being.* Cambridge Massachusetts: Harvard University Press.

Lukes, S. (1996). *The Curious Adventures of Professor Caritat.* London: Verso.

Margalit, A. (2003). *The Ethics of Memory.* Cambridge Massachusetts: Harvard University Press.

Maslow, A. (1954). *Motivation and Personality.* New York: Harper and Row.

Mayr, O. (1986). *Authority, Liberty and Automatic Machinery in Early Modern Europe.* Baltimore: Johns Hopkins University Press.

McConkey, J. (2004). 'Knowledge and Acknowledgement: "Epistemic Injustice" as a Problem of Recognition'. *Politics* 24(3): 198–205.

McCraw, T. (2007). *Prophet of Innovation: Joseph Schumpeter and Creative Destruction.* Cambridge Massachusetts: Harvard University Press.

McGilchrist, I. (2009). *The Master and the Emissary: The Divided Brain and the Making of the Modern World.* New Haven: Yale University Press.

McGonigal, J. (2011). *Reality Is Broken: Why Games Makes Us Better and How They Can Change the World.* New York: Random House.

McKenzie, R. and G. Tullock. (1981). *The New World of Economics: Explorations into the Human Experience.* 3rd edn. (Orig. 1975). Homewood Illinois: Richard D. Irwin.

McLuhan, M. (1965). *Understanding Media: The Extensions of Man.* New York: McGraw-Hill.

Merkle, R. (1992). 'The Technical Feasibility of Cryonics'. *Medical Hypotheses* 39: 6–16.

Mitchell, R. (2010). *BioArt and the Vitality of Media.* Seattle Washington: University of Washington Press.

More, M. (2005). 'The Proactionary Principle'. http://www.maxmore. com/proactionary.htm. Accessed 31 August 2012.

Morrison, G. (2011). *Supergods: Our World in the Age of the Superhero.* London: Jonathan Cape.

DOI: 10.1057/9781137277077

Moss, L. (2003). *What Genes Can't Do*. Cambridge Massachusetts: MIT.

Nadler, S. (2008). *The Best of All Possible Worlds*. New York: Farrar Straus Giroux.

Nagel, T. (1986). *The View from Nowhere*. Oxford: Oxford University Press.

Neiman, S. (2002). *Evil in Modern Thought*. Princeton: Princeton University Press.

Neocleous, M. (2003). 'The Political Economy of the Dead: Marx's Vampires'. *History of the Political Thought* 24: 668–84.

Noble, D.F. (1997). *The Religion of Technology: The Spirit of Invention and the Divinity of Man*. New York: Penguin.

Norman, D. (1988). *The Design of Everyday Things*. New York: Basic Books.

Nussbaum, M. and Sen, A. (eds) (1993). *The Quality of Life*. Oxford: Clarendon Press.

Passmore, J. (1953). *Ralph Cudworth: An Interpretation*. Cambridge UK: Cambridge University Press.

Passmore, J. (1970). *The Perfectibility of Man*. London: Duckworth.

Pichot, A. (2009). *The Pure Society: From Darwin to Hitler*. London: Verso.

Pickering, A. (2010). *The Cybernetic Brain*. Chicago: University of Chicago Press.

Pinker, S. (2002). *The Blank Slate*. New York: Vintage.

Pinker, S. (2008). 'The Stupidity of Dignity'. *The New Republic*. 28 May.

Popper, K. (1957). *The Poverty of Historicism*. London: Routledge and Kegan Paul

Popper, K. (1972). *Objective Knowledge*. Oxford: Oxford University Press.

Popper, K. (1981). 'The Rationality of Scientific Revolutions'. In I. Hacking (ed.), *Scientific Revolutions*. (pp. 80–106) Oxford: Oxford University Press.

Porter, T. (1986). *The Rise of Statistical Thinking, 1820–1900*. Princeton: Princeton University Press.

Proctor, R. (1988) *Racial Hygiene: Medicine under the Nazis*. Cambridge Massachusetts: Harvard University Press.

Proctor, R. (1991). *Value-Free Science? Purity and Power in Modern Knowledge*. Cambridge Massachusetts: Harvard University Press.

Putnam, H. (1982). *Reason, Truth and History*. Cambridge UK: Cambridge University Press.

Rabinbach, A. (1990). *The Human Motor*. New York: Basic Books.

Rasmussen, N. (2008). *On Speed: The Many Lives of Amphetamines*. New York: New York University Press.

DOI: 10.1057/9781137277077

Renwick, C. (2012). *British Sociology's Lost Biological Roots: A History of Futures Past.* London: Palgrave Macmillan.

Rescher, N. (1998). *Predicting the Future.* Albany New York: SUNY Press.

Rescher, N. (1999). *The Limits of Science.* Pittsburgh: University of Pittsburgh Press.

Rifkin, J. (1998). *The Biotech Century.* New York: J.P. Tarcher.

Ripstein, A. (2007). 'As If It Never Happened'. *William & Mary Law Review* 48(5): 1957–97.

Rose, N. (2006). *The Politics of Life Itself: Biomedicine, Power, and Subjectivity in the Twenty-First Century.* Princeton, New Jersey: Princeton University Press.

Rosen, R. (1999). *Essays on Life Itself.* New York: Columbia University Press.

Rothschild, E. (2001). *Economic Sentiments.* Cambridge Massachusetts: Harvard University Press.

Rushkoff, D. (2010). *Program or Be Programmed.* New York: Avalon.

Sarton, G. (1924). 'The New Humanism'. *Isis* 6: 9–24.

Schelling T. (1968). 'The Life You Save May Be Your Own'. In S. Chase (ed.), *Problems in Public Expenditure Analysis.* Washington DC: The Brookings Institution.

Schlick, M. (1974). *The General Theory of Knowledge* (Orig. 1925) Berlin: Springer-Verlag.

Schmitt, C. (1996). *The Concept of the Political.* (Orig. 1932). Chicago: University of Chicago Press.

Schrödinger, E. (1955). *What is Life? The Physical Aspects of the Living Cell* (Orig.1944). Cambridge UK: Cambridge University Press.

Schüklenk, U. and Pacholczyk, A. (2010). 'Dignity's Woolly Uplift'. *Bioethics* 24(2): ii.

Schumpeter, J. (1942). *Capitalism, Socialism and Democracy.* New York: Harper and Row.

Schumpeter, J. (1954). *A History of Economic Analysis.* Oxford: Oxford University Press.

ScienceDaily (2008). 'The Bombardier Beetle, Power Venom and Spray Technologies' (5 April.) http://www.sciencedaily.com/releases/2008/04/080401170543.htm. Accessed 31 August 2012.

Shenk, D. (2010). *The Genius in All of Us.* New York: Doubleday.

Shionoya, Y. (1997). *Schumpeter and the Idea of Social Science.* Cambridge UK: Cambridge University Press.

Simon, H.A. (1977). *The Sciences of the Artificial.* (Orig. 1972). Cambridge Massachusetts: MIT Press.

DOI: 10.1057/9781137277077

Simonton, D. (1984). *Genius, Creativity and Leadership*. Cambridge Massachusetts: Harvard University Press.

Singer, P. (1981). *The Expanding Circle: Ethics and Sociobiology*. New York: Farrar, Strauss and Giroux.

Singer, P. (1993). *Practical Ethics*. Cambridge UK: Cambridge University Press.

Singer, P. (1999). *A Darwinian Left*. London: Weidenfeld and Nicolson.

Smail, D.L. (2008). *On Deep History and the Brain*. Berkeley: University of California Press.

Sørensen, B. (2008). 'Behold, I am Making All Things New: The Entrepreneur as Saviour in the Age of Creativity'. *Scandinavian Journal of Management* 24: 85–93.

Sorgner, S. (2009). 'Nietzsche, the Overman and the Transhuman'. *Journal of Evolution and Technology* 20(1): 29–42.

Stanford, P.K. (2006). *Exceeding Our Grasp*. Oxford: Oxford University Press.

Star, S.L. (1989). *Regions of the Mind: Brain Research and the Quest for Scientific Certainty*. Palo Alto California: Stanford University Press.

Strauss, L. (2000). *On Tyranny*. (Orig. 1948) Chicago: University of Chicago Press.

Taylor, K. (2004). *Brainwashing: The Science of Thought Control*. Oxford: Oxford University Press.

Teilhard de Chardin, P. (1961). *The Phenomenon of Man*. (Orig. 1955) New York: Harper and Row.

Tetlock, P. (2003). 'Thinking the Unthinkable: Sacred Values and Taboo Cognitions'. *Trends in Cognitive Science* 7(7): 320–4.

The Economist (2006). 'Epigenetics: Learning without Learning'. 21 September.

The Economist (2010). 'This House Believes the Development of Computing Was the Most Significant Technological Advance of the 20th Century'. (Online debate, 19–29 October) http://www.economist.com/debate/days/view/598#mod_module. Accessed 31 August 2012.

Toennies, F. (2003). *Critique of Public Opinion*. (Orig. 1922). New York: Walter de Gruyter.

Turkle, S. (1984). *Second Self: Computers and the Human Spirit*. New York: Simon and Schuster.

Turner, S. (2007). 'Social Theory as Cognitive Neuroscience'. *European Journal of Social Theory* 10: 357–74.

Valsiner, J. and van der Veer, R. (2000). *The Social Mind: Construction of the Idea*. Cambridge UK: Cambridge University Press.

DOI: 10.1057/9781137277077

Veit-Brause, I. (2001). 'Scientists and the Cultural Politics of Academic Disciplines in Late 19th Century Germany: Emil Dubois-Reymond and the Controversy Over the Role of the Cultural Sciences'. *History of the Human Sciences* 14(4): 31–56.

Wallerstein, I. (1991). *Unthinking Social Science: The Limits of Nineteenth Century Paradigms.* Oxford: Blackwell.

Weber, M. (1949). *The Methodology of the Social Sciences.* New York: Free Press.

Weber, M. (1958). *The Protestant Ethic and the Spirit of Capitalism.* (Orig. 1904). New York: Scribners.

Wells, H.G. (1938). *World Brain: The Idea of a Permanent Encyclopaedia.* London: Methuen.

White, H. (1972). *Metahistory.* Baltimore: Johns Hopkins University Press.

Wiener, N. (1950). *The Human Use of Human Beings.* Boston: Houghton Mifflin.

Wiener, N. (1964). *God and Golem, Inc.* Cambridge Massachusetts: MIT Press.

Wigner, E. (1960). 'The Unreasonable Effectiveness of Mathematics in the Natural Sciences'. *Communications on Pure and Applied Mathematics* 13(1): 1.14.

Wolbring, G. (2006). 'Ableism and NBICS' (15 August). http://www.innovationwatch.com/choiceisyours/choiceisyours.2006.08.15.htm. Accessed 31 August 2012.

Wood, R. and Orel, V. (2005). 'Scientific breeding in Central Europe in the Early Nineteenth Century: Background to Mendel's Later Work'. *Journal of the History of Biology* 38: 239–72.

Yates, F. (1966). *The Art of Memory.* London: Routledge and Kegan Paul.

Zeleny, J. and Calmes, J. (2008). 'Obama, Assembling Team, Turns to the Economy'. *The New York Times.* 7 November.

DOI: 10.1057/9781137277077

Index

DOI: 10.1057/9781137277077

DOI: 10.1057/9781137277077

DOI: 10.1057/9781137277077

DOI: 10.1057/9781137277077

The manufacturer's authorised representative in the EU is Springer
Nature Customer Service Centre GmbH, Europaplatz 3, 69115 Heidelberg,
Germany. If you have any concerns regarding our products, please
contact ProductSafety@springernature.com

Printed and bound by CPI Group (UK) Ltd, Croydon, CR0 4YY
23/04/2026
02095587-0017